SCIENCE 1105
CHEMICAL FORMULAS, BONDING AND MOLECULAR ARCHITECTURE

CONTENTS

I. CHEMICAL FORMULAS 2
- Ion Charge and Column Location 2
- Formula Prediction 4

II. CHEMICAL BONDING 10
- Electronegativity 10
- Ionic Bonds 12
- Covalent Bonds 13
- Metallic Bonds 16

III. MOLECULAR SHAPE AND ELECTRON DISTRIBUTION 20
- Hybridization 22
- Polarity 29

GLOSSARY 42

Author: **Harold Wengert, Ed.D.**
Editor: Alan Christopherson, M.S.
Illustrations: Juanita Farmer
Stephen Mark Farmer
Kyle Bennett, A.S.

Alpha Omega Publications®

804 N. 2nd Ave. E., Rock Rapids, IA 51246-1759

PERIODIC TABLE

Atomic Properties of the Elements

NIST National Institute of Standards and Technology U.S. Department of Commerce

Physical Measurement Laboratory www.nist.gov/pml

Standard Reference Data www.nist.gov/srd

Frequently used fundamental physical constants

For the most accurate values of these and other constants, visit physics.nist.gov/constants

1 second = 9 192 631 770 periods of radiation corresponding to the transition between the two hyperfine levels of the ground state of ^{133}Cs

speed of light in vacuum	c	299 792 458 m s^{-1}	(exact)
Planck constant	h	6.626 07 x 10^{-34} J s	($\hbar = h/2\pi$)
elementary charge	e	1.602 177 x 10^{-19} C	
electron mass	m_e	9.109 38 x 10^{-31} kg	
	m_ec^2	0.510 999 MeV	
proton mass	m_p	1.672 622 x 10^{-27} kg	
fine-structure constant	α	1/137.035 999	
Rydberg constant	R_∞	10 973 731.569 m^{-1}	
	$R_\infty c$	3.289 841 960 x 10^{15} Hz	
	$R_\infty hc$	13.605 69 eV	
Boltzmann constant	k	1.380 6 x 10^{-23} J K^{-1}	

Solids / Liquids / Gases / Artificially Prepared

Groups: 1 IA, 2 IIA, 3 IIIB, 4 IVB, 5 VB, 6 VIB, 7 VIIB, 8–10 VIII, 11 IB, 12 IIB, 13 IIIA, 14 IVA, 15 VA, 16 VIA, 17 VIIA, 18 VIIIA. Periods 1–7.

Atomic Number	Symbol	Name	Ground-state Level	Standard Atomic Weight	Ground-state Configuration	Ionization Energy (eV)
1	H	Hydrogen	$^2S_{1/2}$	1.008*	1s	13.5984
2	He	Helium	$^1S_0$	4.002602	1s^2	24.5874
3	Li	Lithium	$^2S_{1/2}$	6.94*	1s^{2}2s	5.3917
4	Be	Beryllium	$^1S_0$	9.012182	1s^{2}2s^2	9.3227
5	B	Boron	$^2P^\circ_{1/2}$	10.81*	1s^{2}2s^{2}2p	8.2980
6	C	Carbon	$^3P_0$	12.011*	1s^{2}2s^{2}2p^2	11.2603
7	N	Nitrogen	$^4S^\circ_{3/2}$	14.007*	1s^{2}2s^{2}2p^3	14.5341
8	O	Oxygen	$^3P_2$	15.999*	1s^{2}2s^{2}2p^4	13.6181
9	F	Fluorine	$^2P^\circ_{3/2}$	18.9984032	1s^{2}2s^{2}2p^5	17.4228
10	Ne	Neon	$^1S_0$	20.1797	1s^{2}2s^{2}2p^6	21.5645
11	Na	Sodium	$^2S_{1/2}$	22.98976928	[Ne]3s	5.1391
12	Mg	Magnesium	$^1S_0$	24.3050	[Ne]3s^2	7.6462
13	Al	Aluminum	$^2P^\circ_{1/2}$	26.9815386	[Ne]3s^{2}3p	5.9858
14	Si	Silicon	$^3P_0$	28.085*	[Ne]3s^{2}3p^2	8.1517
15	P	Phosphorus	$^4S^\circ_{3/2}$	30.973762	[Ne]3s^{2}3p^3	10.4867
16	S	Sulfur	$^3P_2$	32.06*	[Ne]3s^{2}3p^4	10.3600
17	Cl	Chlorine	$^2P^\circ_{3/2}$	35.45*	[Ne]3s^{2}3p^5	12.9676
18	Ar	Argon	$^1S_0$	39.948	[Ne]3s^{2}3p^6	15.7596
19	K	Potassium	$^2S_{1/2}$	39.0983	[Ar]4s	4.3407
20	Ca	Calcium	$^1S_0$	40.078	[Ar]4s^2	6.1132
21	Sc	Scandium	$^2D_{3/2}$	44.955912	[Ar]3d4s^2	6.5615
22	Ti	Titanium	$^3F_2$	47.867	[Ar]3d^{2}4s^2	6.8281
23	V	Vanadium	$^4F_{3/2}$	50.9415	[Ar]3d^{3}4s^2	6.7462
24	Cr	Chromium	$^7S_3$	51.9961	[Ar]3d^{5}4s	6.7665
25	Mn	Manganese	$^6S_{5/2}$	54.938045	[Ar]3d^{5}4s^2	7.4340
26	Fe	Iron	$^5D_4$	55.845	[Ar]3d^{6}4s^2	7.9025
27	Co	Cobalt	$^4F_{9/2}$	58.933195	[Ar]3d^{7}4s^2	7.8810
28	Ni	Nickel	$^3F_4$	58.6934	[Ar]3d^{8}4s^2	7.6399
29	Cu	Copper	$^2S_{1/2}$	63.546	[Ar]3d^{10}4s	7.7264
30	Zn	Zinc	$^1S_0$	65.38	[Ar]3d^{10}4s^2	9.3942
31	Ga	Gallium	$^2P^\circ_{1/2}$	69.723	[Ar]3d^{10}4s^{2}4p	5.9993
32	Ge	Germanium	$^3P_0$	72.63	[Ar]3d^{10}4s^{2}4p^2	7.8994
33	As	Arsenic	$^4S^\circ_{3/2}$	74.92160	[Ar]3d^{10}4s^{2}4p^3	9.7886
34	Se	Selenium	$^3P_2$	78.96	[Ar]3d^{10}4s^{2}4p^4	9.7524
35	Br	Bromine	$^2P^\circ_{3/2}$	79.904	[Ar]3d^{10}4s^{2}4p^5	11.8138
36	Kr	Krypton	$^1S_0$	83.798	[Ar]3d^{10}4s^{2}4p^6	13.9996
37	Rb	Rubidium	$^2S_{1/2}$	85.4678	[Kr]5s	4.1771
38	Sr	Strontium	$^1S_0$	87.62	[Kr]5s^2	5.6949
39	Y	Yttrium	$^2D_{3/2}$	88.90585	[Kr]4d5s^2	6.2173
40	Zr	Zirconium	$^3F_2$	91.224	[Kr]4d^{2}5s^2	6.6339
41	Nb	Niobium	$^6D_{1/2}$	92.90638	[Kr]4d^{4}5s	6.7589
42	Mo	Molybdenum	$^7S_3$	95.96	[Kr]4d^{5}5s	7.0924
43	Tc	Technetium	$^6S_{5/2}$	(98)	[Kr]4d^{5}5s^2	7.1194
44	Ru	Ruthenium	$^5F_5$	101.07	[Kr]4d^{7}5s	7.3605
45	Rh	Rhodium	$^4F_{9/2}$	102.90550	[Kr]4d^{8}5s	7.4589
46	Pd	Palladium	$^1S_0$	106.42	[Kr]4d^{10}	8.3369
47	Ag	Silver	$^2S_{1/2}$	107.8682	[Kr]4d^{10}5s	7.5762
48	Cd	Cadmium	$^1S_0$	112.411	[Kr]4d^{10}5s^2	8.9938
49	In	Indium	$^2P^\circ_{1/2}$	114.818	[Kr]4d^{10}5s^{2}5p	5.7864
50	Sn	Tin	$^3P_0$	118.710	[Kr]4d^{10}5s^{2}5p^2	7.3439
51	Sb	Antimony	$^4S^\circ_{3/2}$	121.760	[Kr]4d^{10}5s^{2}5p^3	8.6084
52	Te	Tellurium	$^3P_2$	127.60	[Kr]4d^{10}5s^{2}5p^4	9.0097
53	I	Iodine	$^2P^\circ_{3/2}$	126.90447	[Kr]4d^{10}5s^{2}5p^5	10.4513
54	Xe	Xenon	$^1S_0$	131.293	[Kr]4d^{10}5s^{2}5p^6	12.1298
55	Cs	Cesium	$^2S_{1/2}$	132.9054519	[Xe]6s	3.8939
56	Ba	Barium	$^1S_0$	137.327	[Xe]6s^2	5.2117
72	Hf	Hafnium	$^3F_2$	178.49	[Xe]4f^{14}5d^{2}6s^2	6.8251
73	Ta	Tantalum	$^4F_{3/2}$	180.94788	[Xe]4f^{14}5d^{3}6s^2	7.5496
74	W	Tungsten	$^5D_0$	183.84	[Xe]4f^{14}5d^{4}6s^2	7.8640
75	Re	Rhenium	$^6S_{5/2}$	186.207	[Xe]4f^{14}5d^{5}6s^2	7.8335
76	Os	Osmium	$^5D_4$	190.23	[Xe]4f^{14}5d^{6}6s^2	8.4382
77	Ir	Iridium	$^4F_{9/2}$	192.217	[Xe]4f^{14}5d^{7}6s^2	8.9670
78	Pt	Platinum	$^3D_3$	195.084	[Xe]4f^{14}5d^{9}6s	8.9588
79	Au	Gold	$^2S_{1/2}$	196.966569	[Xe]4f^{14}5d^{10}6s	9.2256
80	Hg	Mercury	$^1S_0$	200.59	[Xe]4f^{14}5d^{10}6s^2	10.4375
81	Tl	Thallium	$^2P^\circ_{1/2}$	204.38*	[Hg]6p	6.1083
82	Pb	Lead	$^3P_0$	207.2	[Hg]6p^2	7.4167
83	Bi	Bismuth	$^4S^\circ_{3/2}$	208.98040	[Hg]6p^3	7.2855
84	Po	Polonium	$^3P_2$	(209)	[Hg]6p^4	8.414
85	At	Astatine	$^2P^\circ_{3/2}$	(210)	[Hg]6p^5	9.350
86	Rn	Radon	$^1S_0$	(222)	[Hg]6p^6	10.7485
87	Fr	Francium	$^2S_{1/2}$	(223)	[Rn]7s	4.0727
88	Ra	Radium	$^1S_0$	(226)	[Rn]7s^2	5.2784
104	Rf	Rutherfordium	$^3F_2$	(265)	[Rn]5f^{14}6d^{2}7s^2	6.01
105	Db	Dubnium	$^4F_{3/2}$	(268)	[Rn]5f^{14}6d^{3}7s^2	6.8
106	Sg	Seaborgium		(271)	[Rn]5f^{14}6d^{4}7s^2	7.8
107	Bh	Bohrium		(270)	[Rn]5f^{14}6d^{5}7s^2	7.7
108	Hs	Hassium		(277)	[Rn]5f^{14}6d^{6}7s^2	7.6
109	Mt	Meitnerium		(276)		
110	Ds	Darmstadtium		(281)		
111	Rg	Roentgenium		(280)		
112	Cn	Copernicium		(285)		
113	Uut	Ununtrium		(284)		
114	Fl	Flerovium		(289)		
115	Uup	Ununpentium		(288)		
116	Lv	Livermorium		(293)		
117	Uus	Ununseptium		(294)		
118	Uuo	Ununoctium		(294)		

Lanthanides

Atomic Number	Symbol	Name	Ground-state Level	Standard Atomic Weight	Ground-state Configuration	Ionization Energy (eV)
57	La	Lanthanum	$^2D_{3/2}$	138.90547	[Xe]5d6s^2	5.5769
58	Ce	Cerium	$^1G^\circ_4$	140.116	[Xe]4f5d6s^2	5.5386
59	Pr	Praseodymium	$^4I^\circ_{9/2}$	140.90765	[Xe]4f^{3}6s^2	5.473
60	Nd	Neodymium	$^5I_4$	144.242	[Xe]4f^{4}6s^2	5.5250
61	Pm	Promethium	$^6H^\circ_{5/2}$	(145)	[Xe]4f^{5}6s^2	5.582
62	Sm	Samarium	$^7F_0$	150.36	[Xe]4f^{6}6s^2	5.6437
63	Eu	Europium	$^8S^\circ_{7/2}$	151.964	[Xe]4f^{7}6s^2	5.6704
64	Gd	Gadolinium	$^9D^\circ_2$	157.25	[Xe]4f^{7}5d6s^2	6.1498
65	Tb	Terbium	$^6H^\circ_{15/2}$	158.92535	[Xe]4f^{9}6s^2	5.8638
66	Dy	Dysprosium	$^5I_8$	162.500	[Xe]4f^{10}6s^2	5.9391
67	Ho	Holmium	$^4I^\circ_{15/2}$	164.93032	[Xe]4f^{11}6s^2	6.0215
68	Er	Erbium	$^3H_6$	167.259	[Xe]4f^{12}6s^2	6.1077
69	Tm	Thulium	$^2F^\circ_{7/2}$	168.93421	[Xe]4f^{13}6s^2	6.1843
70	Yb	Ytterbium	$^1S_0$	173.054	[Xe]4f^{14}6s^2	6.2542
71	Lu	Lutetium	$^2D_{3/2}$	174.9668	[Xe]4f^{14}5d6s^2	5.4259

Actinides

Atomic Number	Symbol	Name	Ground-state Level	Standard Atomic Weight	Ground-state Configuration	Ionization Energy (eV)
89	Ac	Actinium	$^2D_{3/2}$	(227)	[Rn]6d7s^2	5.3802
90	Th	Thorium	$^3F_2$	232.03806	[Rn]6d^{2}7s^2	6.3067
91	Pa	Protactinium	$^4K_{11/2}$	231.03588	[Rn]5f^{2}6d7s^2	5.89
92	U	Uranium	$^5L^\circ_6$	238.02891	[Rn]5f^{3}6d7s^2	6.1941
93	Np	Neptunium	$^6L_{11/2}$	(237)	[Rn]5f^{4}6d7s^2	6.2655
94	Pu	Plutonium	$^7F_0$	(244)	[Rn]5f^{6}7s^2	6.0258
95	Am	Americium	$^8S^\circ_{7/2}$	(243)	[Rn]5f^{7}7s^2	5.9738
96	Cm	Curium	$^9D^\circ_2$	(247)	[Rn]5f^{7}6d7s^2	5.9914
97	Bk	Berkelium	$^6H^\circ_{15/2}$	(247)	[Rn]5f^{9}7s^2	6.1978
98	Cf	Californium	$^5I_8$	(251)	[Rn]5f^{10}7s^2	6.2817
99	Es	Einsteinium	$^4I^\circ_{15/2}$	(252)	[Rn]5f^{11}7s^2	6.3676
100	Fm	Fermium	$^3H_6$	(257)	[Rn]5f^{12}7s^2	6.50
101	Md	Mendelevium	$^2F^\circ_{7/2}$	(258)	[Rn]5f^{13}7s^2	6.58
102	No	Nobelium	$^1S_0$	(259)	[Rn]5f^{14}7s^2	6.65
103	Lr	Lawrencium	$^2P^\circ_{1/2}$	(262)	[Rn]5f^{14}7s^{2}7p	4.90

Key: Atomic Number (58), Ground-state Level ($^1G^\circ_4$), Symbol (Ce), Name (Cerium), Standard Atomic Weight† (140.116), Ground-state Configuration ([Xe]4f5d6s^2), Ionization Energy (eV) (5.5386)

†Based upon ^{12}C. () indicates the mass number of the longest-lived isotope.

*IUPAC conventional atomic weights; standard atomic weights for these elements are expressed in intervals; see iupac.org for an explanation and values.

For a description of the data, visit physics.nist.gov/data

NIST SP 966 (March 2013)

SCIENCE 1105
CHEMICAL FORMULAS, BONDING AND MOLECULAR ARCHITECTURE

So far you have studied measurement, classification and properties of matter, and atomic structure. All of the ideas and concepts you have learned in chemistry are necessary to understand the chemistry you will study in the rest of this course. The perfect order of our Creator makes His creation predictable. Scientists will never unlock all of the secrets of nature, but some of the regularities and predictions man has learned will be studied in this LIFEPAC®.

OBJECTIVES

Read these objectives. The objectives tell you what you will be able to do when you have successfully completed this LIFEPAC.

When you have finished this LIFEPAC, you should be able to:

1. Predict chemical formulas by using the Periodic Table.
2. Define and use the concepts of electronegativity.
3. Predict percent ionic and covalent character of bonds.
4. Describe and explain the nature of ionic, covalent, and metallic bonds.
5. Describe and explain the concept of hybridization.
6. Predict molecular shape by knowing the molecular formula.
7. Describe polarity and predict the polarity in molecules.

Survey the LIFEPAC. Ask yourself some questions about this study. Write your questions here.

I. CHEMICAL FORMULAS

How do elements combine? Why do some elements combine in multiple numbers like H_2O, Na_2SO_4, $KMnO_4$, and others combine in one-to-one ratios like NaCl, NaI, CaO, and AgCl? Why is the formula for water H_2O rather than HO or HO_2? Science LIFEPAC 1104 should help give the answers. Have Science LIFEPAC 1104 handy because it will be used often. Also, have the Periodic Table readily available.

SECTION OBJECTIVE

Review this objective. When you have completed this section, you should be able to:

1. Predict chemical formulas by using the Periodic Table.

VOCABULARY

Study this word to enhance your learning success in this section.

ion

Note: All vocabulary words in this LIFEPAC appear in **boldface** print the first time they are used. If you are unsure of the meaning when you are reading, study the definitions given.

ION CHARGE AND COLUMN LOCATION

You may recall from Science LIFEPAC 1104 that the valence electrons are the electrons that determine the chemistry of an atom. Each element of each column had the same valence electron structure as each other element of that column. Because of the similarity among elements of each column, the chemistry of all members of a column (called a *family*) is similar. Therefore, if we know the chemical properties of one member of the family, we know the properties of all members of that family.

Experimental evidence. The following experiment was carried out in a laboratory setting. You will not do it yourself because the chemicals are too expensive. You will be expected to analyze the results, however.

Solutions of sodium chloride, magnesium chloride, and aluminum chloride were made so that each had 0.10 moles of compound per liter of solution.

Do these activities.

1.1 How many moles of each chloride are in one milliliter of the respective solutions?

____

1.2 Assume that you measured exactly 2.00 ml of the sodium chloride into a small test tube. How many moles of sodium chloride are in the tube? __________________

1.3 Consider the situation if your measurements were not exact and you got 0.01 ml too much solution.

a. How many moles extra NaCl would you have? __________________

b. What percent error would this excess produce? __________________

1.4 Assume you measure out exactly 2.00 ml of magnesium chloride into a second test tube. How many moles of magnesium chloride are in the tube? __________________

1.5 Now assume you measure out exactly 2.00 ml of aluminum chloride into a third test tube. How many moles of aluminum chloride are in the tube? __________________

A silver nitrate solution that was four times more concentrated than the chloride solutions was added to the chloride. Generally, when silver collides with the chloride they react and form a solid, and a *precipitate* is formed. The 4:1 molar ratio will insure that more silvers than chlorides cause all chlorides to precipitate out. A white precipitate formed in each test tube, and the test tubes were let stand about five minutes. The height of the precipitate in the test tube was measured and recorded. Using the height of the precipitate in the sodium chloride test tube as a standard equal to 1, the magnesium chloride precipitate was two times as high. Again with the sodium chloride as the standard of 1, the aluminum chloride precipitate was three times as high. Silver chloride was the precipitate in each reaction.

Do these activities.

1.6 The mole ratio of silver chloride produced in each case using sodium chloride as the reference was 1: _______ : _______

1.7 In this experiment what is the whole-number mole ratio of sodium chloride to magnesium chloride to aluminum chloride? _______ : _______ : _______

1.8 Using the whole number ratio in 1.7 and the ratio for AgCl produced and calculated in 1.6, complete balancing the equations for three reactions.

a. _______ $NaCl$ + _______ $AgNO_3$ $\longrightarrow$ _______ $AgCl$ + _______ $NaNO_3$

b. _______ $MgCl_2$ + _______ $AgNO_3$ $\longrightarrow$ _______ $AgCl$ + _______ $Mg(NO_3)_2$

c. _______ $AlCl_3$ + _______ $AgNO_3$ $\longrightarrow$ _______ $AgCl$ + _______ $Al(NO_3)_3$

1.9 For some reason one sodium atom combines with one chlorine atom (NaCl), one magnesium combines with two chlorine atoms ($MgCl_2$), and one aluminum atom combines with three chlorine atoms ($AlCl_3$). In Science LIFEPAC 1104 you learned that valence electron structure was the key to the chemistry of the various atoms. Why do the three metals each combine differently with chlorine? (Review Science LIFEPAC 1104).

ELEMENT	Na	Mg	Al	Cl	INERT GAS
Family Member	IA	a.	b.	c.	d.
Valence e- Number	e.	2	f.	g.	h.
Family Valence e- Number	i.	j.	k.	7	l.

1.10 Answer these questions about the previous chart.

a. What is the most stable family of elements? _______

b. How many valence electrons does this family have? _______

c. What is true about the *s* and *p* sublevels of each member? _______________________

____

Valence electron charge. Nearly all atoms tend to gain or lose valence electrons in a chemical reaction to take on the inert gas structure and have complete *s* — *p* sublevels. Whether a family of elements gains or loses depends upon which is easier. For example, less energy is required for sodium to lose its one valence electron and take on the electron structure of Ne than to gain seven electrons to have the structure of Ar.

Do these activities.

1.11 How would Mg change to gain an inert gas structure?

____

____

1.12 How would Al change to gain stability?

____

____

1.13 How could Cl change to gain stability?

____

____

Adult check ______________
Initial Date

FORMULA PREDICTION

When the sodium atom, magnesium atom, aluminum atom, and chlorine atom change their outer electron shells to a more stable configuration, they become **ions**. This change means that they have gained or lost electrons and now have an unequal number of electrons (-) and protons (+). Since in a chemical reaction the atomic number [number of protons (+)] *always* remains the same, a gain or loss of electrons will result in the particle having a net electrical charge. An example might be lithium. A lithium atom enters a chemical reaction with three electrons and three protons, but generally loses one electron. This loss means that lithium has a net charge of +1, meaning one excess positive charge or one more proton than electron.

Before: 3+ and 3- = net charge of 0
After: 3+ and 2- = net charge of +1 (lost one electron)

Complete the following table using the ideas from the previous paragraph.

	Elements	Electron Structure of Atom	Electron Structure of Ion	Net Ion Charge
	Lithium	$1s^2\ 2s^1$	$1s^2$	1+
1.14	Beryllium	a.	b.	c.
1.15	Chlorine	a.	b.	c.
1.16	Sodium	a.	b.	c.
1.17	Fluorine	a.	b.	c.
1.18	Oxygen	a.	b.	c.
1.19	Magnesium	a.	b.	c.
1.20	Calcium	a.	b.	c.
1.21	Boron	a.	b.	c.
1.22	Phosphorous	a.	b.	c.
1.23	Sulfur	a.	b.	c.
1.24	Aluminum	a.	b.	c.
1.25	Nitrogen	a.	b.	c.
1.26	Silicon	a.	b.	c.
1.27	Carbon	a.	b.	c.

Adult check ______________________
Initial Date

Do these activities.

1.28 Locate on the Periodic Table each group of ions with the same charge. What relationship exists between ions of the same charge and the location of the corresponding element on the Periodic Table? ____
____
____

1.29 From the relationship in 1.28 complete the table for ion charge based upon their periodic location. (Use the Periodic Table as necessary.)

Column	Ion Charge	Number of Valence Electrons
IA	a. ______________	b. ______________
IIA	c. ______________	d. ______________
IIIA	e. ______________	f. ______________
IVA	g. ______________	h. ______________
VA	i. ______________	j. ______________
VIA	k. ______________	l. ______________
VIIA	m. ______________	n. ______________
VIIIA	o. ______________	p. ______________

1.30 When ions combine, the product is electrically neutral. This process means that the number of +'s will equal the number of -'s. For example, A^{+1} combines with E^{-1} in a 1:1 ratio, AE. The ion A^{+1} combines with G^{-2} in a 2:1 ratio to be neutral and the formula would be A_2G.

a. If D^{+2} would react with E^{-1}, what do you predict to be the formula? ______________

b. Predict the formula for D^{+2} and G^{-2}. ______________

c. Would A^{+1} and D^{+2} react? ______________

d. Explain your reasoning for the predicted formula in c. above.

____
____

Predict the following chemical formulas.

	Compound	Predicted Formula	Check a Chemistry Handbook for the Accepted Formula
1.31	Potassium Bromide		
1.32	Sodium Chloride		
1.33	Calcium Chloride		
1.34	Lithium Fluoride		
1.35	Cesium Sulfide		
1.36	Strontium Bromide		
1.37	Magnesium Oxide		
1.38	Beryllium Sulfide		
1.39	Rubidium Chloride		
1.40	Barium Oxide		
1.41	Aluminum Chloride		
1.42	Aluminum Oxide		
1.43	Magnesium Chloride		

1.44 Refer to 1.29 to complete these activities.

a. Do the positive ions tend to be metals or nonmetals? ____________________

b. Do the negative ions tend to be metals or nonmetals? ____________________

c. Based on these ideas, make a general rule relating ion charge to periodic location.

____

____

Adult check ____________________
Initial Date

1.45 **OPTIONAL:** Predict the formulas of the following compounds.

a. magnesium sulfide = ________________

b. aluminum iodide = ________________

c. cesium oxide = ________________

d. phosphorous tri-iodide = ________________

e. hydrogen oxide = ________________

f. carbon tetrachloride = ________________

g. silicon dioxide = ________________

h. aluminum oxide = ________________

1.46 **OPTIONAL:** The following list contains some common ions. Using the charge on these ions and the idea of valence, predict the formulas for the following compounds.

nitrate	= NO_3^-	phosphate	= PO_4^{-3}
sulfate	= SO_4^{-2}	acetate	= CH_3COO^-
ammonium	= NH_4^+	chromate	= CrO_4^{-2}
carbonate	= CO_3^{-2}	dichromate	= $Cr_2O_7^{-2}$
permanganate	= MnO_4^-	sulfide	= S^{-2}

a. sodium carbonate = ______________

b. calcium phosphate = ______________

c. hydrogen sulfide = ______________

d. potassium permanganate = ______________

e. calcium carbonate = ______________

f. aluminum nitrate = ______________

g. ammonium dichromate = ______________

h. barium sulfate = ______________

i. lithium chromate = ______________

j. magnesium acetate = ______________

k. ammonium carbonate = ______________

l. strontium sulfide = ______________

1.47 **OPTIONAL:** Balance the following reactions by completing the mole ratios.

a. $Ca_3(PO_4)_2$ + ______ C + $3SiO_2$ → ______ $CaSiO_3$ + ______ P + ______ CO

b. ______ P + ______ O_2 → $2P_2O_5$

c. ______ As_2S_3 + $9O_2$ → $2As_2O_3$ + ______ SO_2

d. CS_2 + ______ Cl_2 → CCl_4 + S_2Cl_2

e. ______ Na_2CO_3 + $CaCO_3$ + 6 SiO_2 → $Na_4CaSi_6O_{15}$ + ______ CO_2

f. ______ C_8H_{18} + $25O_2$ → 16 CO_2 + ______ H_2O

g. $2CH_3OH$ + ______ O_2 → $2CO_2$ + ______ H_2O

h. $2C_2H_2$ + ______ O_2 → $4CO_2$ + ______ H_2O

Review the material in this section in preparation for the Self Test. The Self Test will check your mastery of this particular section. The items missed on this Self Test will indicate specific areas where restudy is needed for mastery.

Use the Periodic Table as necessary for this test.

Complete the following activities (each answer, 3 points).

1.01 An element of Column IIA loses two electrons in the process of a chemical combination. What is its ion charge? ______________

1.02 By using the Periodic Table determine the ion charges of the following families of elements.

a. Column IA = ______________

b. Column Oxygen = ______________

c. Column IIIA = ______________

d. Column Barium = ______________

e. Column Iodine = ______________

f. Column VIIIA = ______________

Circle the letter of the best response (each answer, 3 points).

1.03 Which one of the following elements will combine with oxygen to give an M_2O_3 formula where the unknown element to oxygen ratio is 2:3?

a. Na b. Be c. Al d. Si e. Cl

1.04 Which one of the following elements will combine with chlorine in a 1:2 ratio to give a formula of MCl_2?

a. Na b. Al c. Mg d. F e. C

1.05 Which of the following formulas best represents the compound of carbon and bromine?

a. C_4Br b. C_2Br_3 c. CBr d. C_2Br_2 e. CBr_4

1.06 How many moles of Br will be needed to completely combine with 0.01 moles of C?

a. 0.02 b. 0.04 c. 0.4 d. 0.2 e. cannot tell

1.07 How many moles of Cl^- ions are needed to completely combine with 0.25 moles of Mg^{+2} ions?

a. 0.175 b. 0.75 c. 0.50 d. 0.25 e. 0.125

1.08 Neutral atoms of neon, atomic number of 10, have the same number of electrons as each of the following except

a. Na^+ b. Mg^{+2} c. F^- d. Ar e. O^{-2}

1.09 The most likely formula for the combination of Sr with Br is

a. $SrBr_2$ b. SrBr c. Sr_2Br d. $SrBr_3$ e. Sr_2Br_3

1.010 Based upon the location of Column IVA on the Periodic Table, the ion charge should be

a. +1 b. +2 c. +3 d. ±4 e. -3

1.011 Which of the following elements will form a compound with iodine to give a formula of the form MI_3 where three atoms of iodine react with one of the elements?

a. Al b. Na c. C d. Be e. Ag

1.012 Neutral atoms of argon, atomic number 18, have the same number of electrons as each of the following items *except*

a. K^+ b. Ca^{+2} c. Ne d. Cl^- e. S^{-2}

Do the following items (each answer, 3 points).

1.013 Based upon the ion charge of the following groups of atoms that react together as a group, predict the formula for the following compounds.

sulfate = SO_4^{-2} phosphate = PO_4^{-3} hydroxide = OH^-

a. calcium hydroxide = __________

b. sodium sulfate = __________

c. aluminum phosphate = __________

1.014 Based upon the location of the elements in the Periodic Table, what is the net ion charge of the following elements?

a. Al = _____	d. P = _____	g. Cl = _____
b. Na = _____	e. K = _____	h. O = _____
c. Si = _____	f. H = _____	i. He = _____

1.015 If the element sulfur, S, were to lose two electrons per atom, the net ion charge of each atom would be __________ .

72/90

Score __________
Adult check __________
Initial Date

II. CHEMICAL BONDING

One of the major studies in chemistry is concerned with how atoms group together. Why do atoms combine at all? What makes them stick together once they have formed? What forces cause NaCl to stay in a cubic form? This section will help to answer these questions.

SECTION OBJECTIVES

Review these objectives. When you have completed this section, you should be able to:

2. Define and use the concept of electronegativity.
3. Predict percent ionic and covalent character of bonds.
4. Describe and explain the nature of ionic, covalent, and metallic bonds.

VOCABULARY

Study these words to enhance your learning success in this section.

covalent bond | ionic bond | metallic bond
electronegativity

ELECTRONEGATIVITY

Scientists found that all atoms in the combined state show varying degrees of attraction for the electrons of the other member(s) of the compound. For example, the fluorine atom in HF has a very strong attraction for the electron of the hydrogen. In HCl is found a similar but weaker attraction, and in HBr an even weaker attraction is observed. Look at the Periodic Table of the Elements that you found as an insert to Science LIFEPAC 1101. Notice that the atomic size increases as you go down a vertical column. This increase can be represented in this way:

$$F < Cl < Br$$

The molecules of HF, HCl, and HBr could be represented in this way:

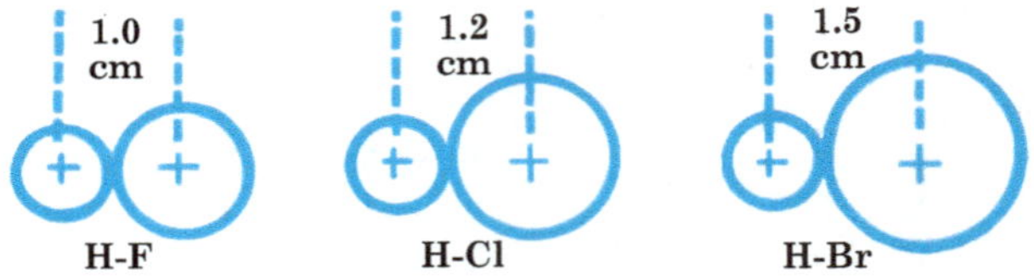

Do these activities.

2.1 Measure the distance between nuclei in each molecule.

a. HCl molecular distance between nuclei = ____________

b. HF molecular distance between nuclei = ____________

c. HBr molecular distance between nuclei = ____________

2.2 Remember that each nucleus attracts the electrons of the other atom to hold the molecule together. Which atom has the greatest distance to overcome to attract the hydrogen electron (Circle the greatest one)?

fluorine | chlorine | bromine

2.3 If two opposite charges get farther apart, does their attraction for each other increase, decrease, or remain the same? ____________

2.4 Which element then has the least attraction for the hydrogen atom, fluorine, chlorine, or bromine?

____

2.5 Explain why HF is a more strongly bonded molecule than HBr.

____

____

____

____

Ionization energy. Review Science LIFEPAC 1104 concerning ionization energy. Remember that ionization energy is the energy necessary to remove valence electrons from atoms. Look carefully at the Periodic Table. Notice that there is a general increase in ionization energy, E_I, from left to right across a periodic row and general decrease in the ionization energy when going down a periodic column.

Increasing ionization energy means that generally *more energy* is required to remove an electron from the elements going *across a row*, and it takes progressively *less energy* to remove an electron when going *down a column*. This progression can be represented in general as in Figure 1.

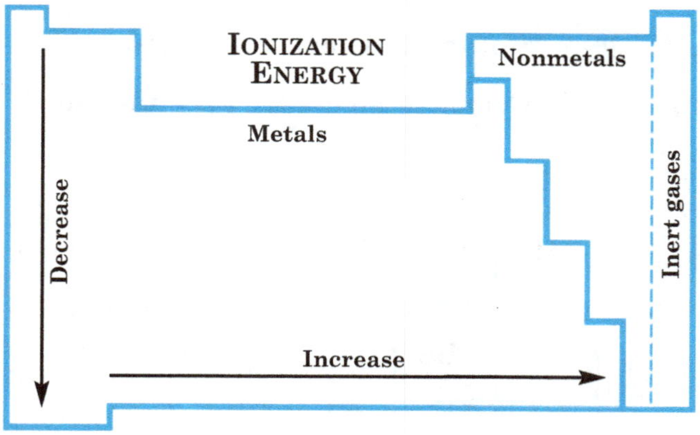

FIGURE 1

Another idea might be drawn from these observations. An atom that gives up an electron easily is less likely to take on another electron. This fact means that the atoms with less ability to hold any given electron (ones with low ionization energy) will not be very likely to want or to get another electron. Atoms with high ionization potential will be able to hold another electron more easily. Therefore, in general, greater ionization energy indicates a greater ability to attract additional electrons; the lower the ionization energy, the less the desire for additional electrons.

Do this activity.

2.6 On the basis of the preceding reasoning, answer these questions.

a. Do metals or nonmetals have the greatest chance to lose electrons and become positive ions?

____

b. Which group, metals or nonmetals, has the greatest chance to gain electrons and to become negative ions? ______________________________

c. Which group of elements has very little desire to gain or to lose electrons?

____

Adult check ____________________
Initial Date

Definition. Refer to the Periodic Table, which shows the electron *attracting ability*, or **electronegativity** as it is more commonly called. Pauling developed a scale of relative values with fluorine being assigned a value of 4 and all other elements rated on that basis. Complete Figure 2 in 2.7 by putting in the corresponding electronegativities for each element listed.

Complete these activities.

2.7 Complete Figure 2 by putting the electronegativities of each element on the line in the box.

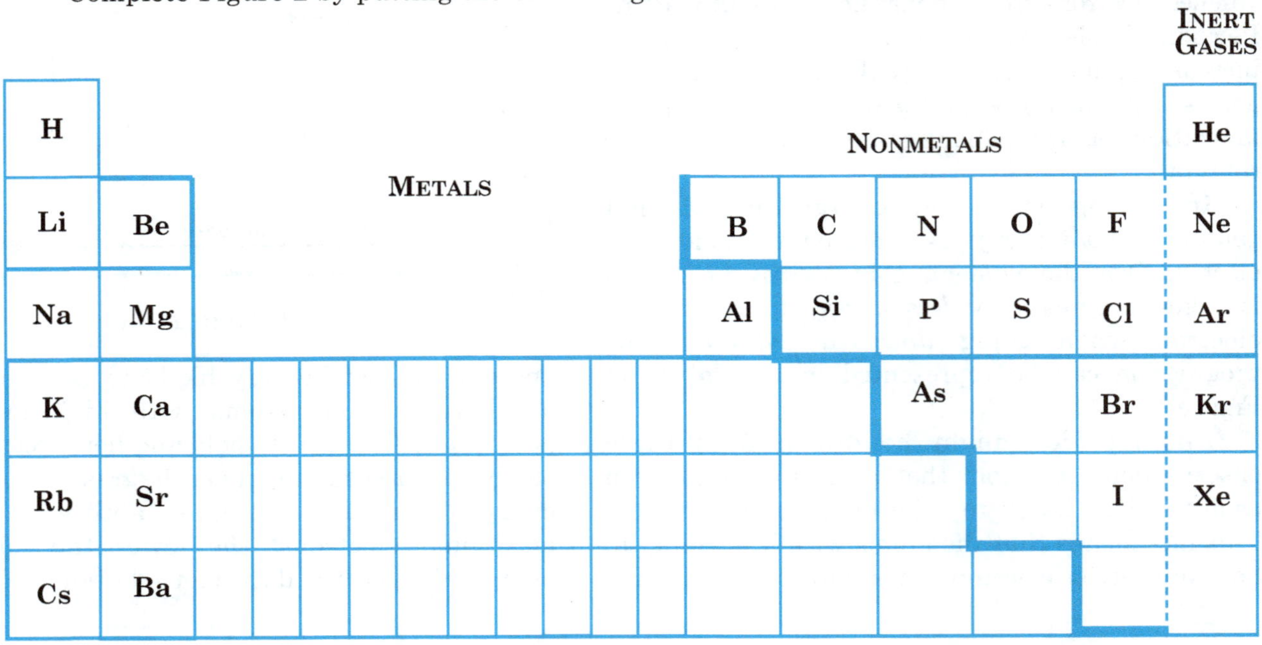

FIGURE 2

2.8 In relationship to Figure 1 for ionization energy, compare in general how ionization energy and electronegativity are related to the Periodic Table.

____

____

____

____

____

IONIC BONDS

When elements with different abilities to attract the electrons of another element are put together, a bond will result. If one element can completely take a valence electron away from some other atom, ions will form.

Definition. An **ionic bond** is *any bond that is made from ions.* This relationship is shown in this diagram.

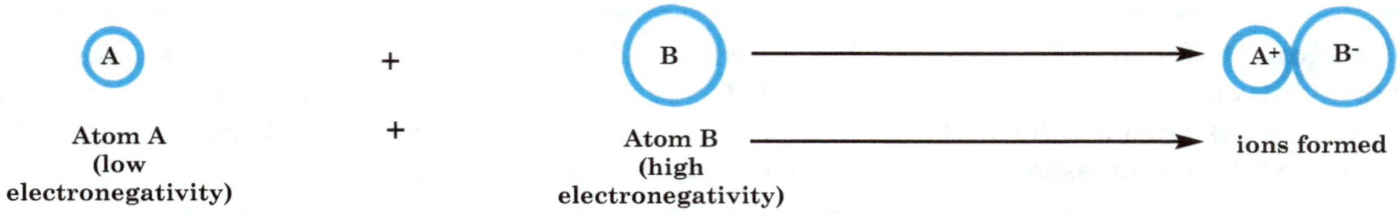

Complete this question.

2.9 Why does an ionic bond stay together? (Hint: Opposite electrical charges.)

Characteristics. Ionic bonds are the most common bonds in the minerals of the earth. Most crystals are ionic in nature. Some common ionic compounds include table salt (NaCl), lithium chloride (LiCl), and calcium chloride ($CaCl_2$). The ionic bond involves a transfer of a valence electron from one atom to another. This transfer is shown in this diagram.

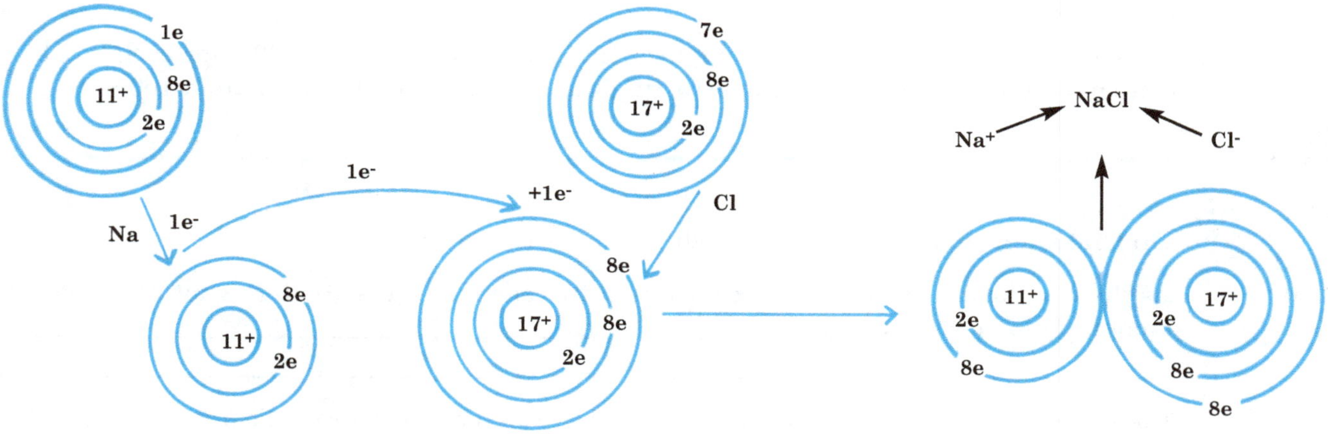

COVALENT BONDS

Most ionic bonds occur between metals and nonmetals. Nearly all inorganic substances are ionic in nature. A few notable exceptions are water (H_2O), carbon dioxide (CO_2), and atmospheric gases like oxygen (O_2), hydrogen (H_2), and nitrogen (N_2). These exceptions are a part of a second class of bonds called **covalent bonds**.

Definition. Sometimes a bond will form between two or more atoms that have similar electronegativities. In this case, Atom A has no more attraction for the electrons of Atom B than Atom B has for the electrons of Atom A. Each has a mutual attraction for the other's valence electrons. When this sharing bond occurs, the resulting bond is called a covalent bond. This *covalent bond* is shown in this diagram.

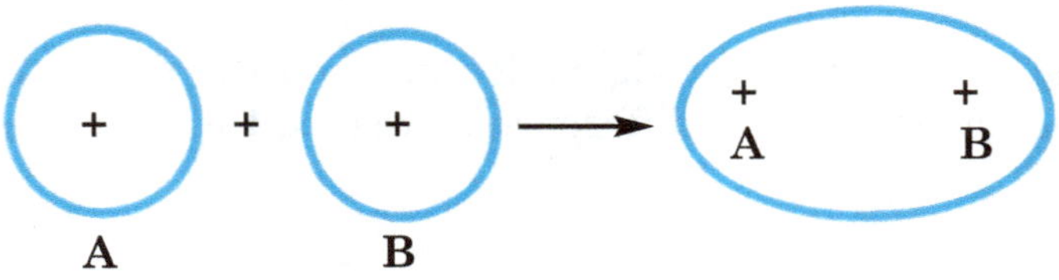

In a covalent bond, two or more electrons are located in the region between the nuclei. This bond is a highly directed bond because the bonding electrons are located only in specific places.

Percent ionic character. Chemists find that most bonds are partly ionic and partly covalent. The amount of ionic and covalent character is determined by the difference in the electronegativities of the atoms making up the compound. The mathematics resulting in a prediction device for determining the degree of ionic character of any bond was studied by several men. We will use the chart as developed by Linus Pauling, a two time Nobel Prize winner. Refer to the Periodic Table found in Science LIFEPAC 1101. In the upper center of the Periodic Table is a chart that displays the *Difference in electronegativity* and the *Percent ionic character*. This chart is based on fluorine, which has the greatest attraction for electrons and is given a value of 4. All of the other elements have a value relative to the value for fluorine.

You can determine the percent of ionic character of a single bond by this method:

1. Calculate the numerical difference in the electronegativity values of the two atoms involved in the bond.
2. Locate this value in the top row of the chart labeled "Difference in electronegativity."
3. Read the value directly below this electronegativity difference. This second value is the percent of ionic character.

Example. Determine the ionic character of the bond between H and F in the molecule HF.

	Element	**Electronegativity**
Step 1:	F	4.0
	H	2.1
	difference =	1.9

Step 2: Find 1.9 in the top row of the chart.

Step 3: Locate the percent directly below 1.9. The percent ionic character is equal to 59%.

The total percent ionic character and percent covalent character will equal 100% for each bond. The percent covalent character is determined by subtracting the percent ionic character from 100%.

Do these activities. Use the Periodic Table as necessary.

2.10 As the electronegativity differences between two elements gets greater, what happens to the amount of ionic character? ____

____

2.11 Does this greater difference mean that the resulting bond is more ionic or covalent?

____

2.12 Answer these questions about electronegativity difference.

a. If the bond were 50% ionic and 50% covalent, what would be the electronegativity difference? ____________________

____

b. If the electronegativity difference were 1.0 what would be the percent ionic character?

____

c. Does a 1.0 electronegativity difference represent a predominantly ionic or covalent type bond?

____

2.13 a. Which compound would have a more ionic bond, NaCl or NaI? ____________________

b. Which compound has a greater difference in electronegativities of its members, NaCl or KCl?

____

2.14 Refer to the first paragraph following the section title "COVALENT BONDS." List the three substances that are 100% covalent in nature. a. ______________ b. ______________ c. ______________

2.15 Complete the following generalization about electronegativity difference and the degree of ionic character of a chemical bond. As electronegativity difference increases, the degree of ionic character ______________ .

2.16 Find the approximate percent of ionic character of the following compounds. Consider that each is bonded only between two elements with a single bond. For example, the electronegativity difference for H_2S is 0.4, *not* (0.4) (2). Use a chemistry handbook to find whether each is a solid, liquid, or gas at room temperature.

NaCl	= a. ______ ;	b. ______
CsF	= c. ______ ;	d. ______
$MgCl_2$	= e. ______ ;	f. ______
BrF_5	= g. ______ ;	h. ______
CO_2	= i. ______ ;	i. ______
H_2O	= k. ______ ;	l. ______
O_2	= m. ______ ;	n. ______
NO_2	= o. ______ ;	p. ______
IF_5	= q. ______ ;	r. ______

2.17 Consider these questions.

a. Generally, how does the amount of ionic character relate to the state of matter?

____

____

b. How does bond strength relate to ionic character? ______________________

____

c. How does melting point and boiling point relate to bond type? ______________

____

2.18 Bonding between metals and nonmetals is common in the minerals of the earth.

a. In general when a metal bonds with a nonmetal, does the bond type tend to be more ionic or covalent? ______________

b. Predict what you think is the most ionic compound possible. The formula is __________ .

c. Why did you predict this combination? ______________________

____

____

____

2.19 We found examples in 2.14 of 100% covalent bonds.

a. Can you ever have 100% ionic compound? ______

b. What is the basis of this prediction? ______________________

____

____

2.20 Which two columns of elements when combined with each other tend to form the most ionic compounds? ____

____

METALLIC BONDS

The third type of bonding that we will study in some detail is that type which holds metals together. All metals except mercury are solids at room temperature. We also know that all metals are good conductors of electricity and heat.

Definition. Let us investigate conductivity and other metallic bonding characteristics to expand your ideas about metals. Look at the three areas generally outlined on the Periodic Table in Figure 3 as inert gases, metals, or nonmetals. **Metallic bonds** are bonds that hold metals together.

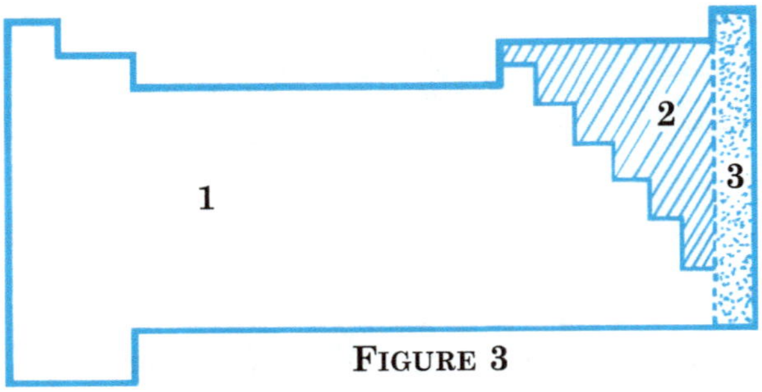

FIGURE 3

1. ____________________
2. ____________________
3. ____________________

Do these activities.

2.21 Label the three areas in Figure 3 as metals, inert gases, or nonmetals.

2.22 Consider these questions.

a. Generally, which of the three families of elements listed in 2.21 has the greatest ionization energy? ____________________

b. Which family generally has the valence electrons which are the easiest to remove from the atom? ____________________

c. Therefore, which family of elements will have the most mobile electrons? ____________

Characteristics. You have already investigated several properties of metals. They are good conductors of heat and electricity, are generally solids at room temperature, and are usually shiny in color. Most metals are malleable (capable of being shaped or hammered) and do not come in a pure form in their natural state. Gold and silver are exceptions to the pure-form rule.

The metallic bond is different from the ionic or covalent bond. The ionic bonds and covalent bonds are both *directed bonds*. Directed bonds means the bonding electrons are located between the two atoms of the bond which causes the bonding electrons to be "locked" into a fixed location. Metallic bonds are made from a "sea" of electrons and the bonding electrons are free to roam among any of the atoms in the metal piece. A diagram of a metallic bond is shown in Figure 4.

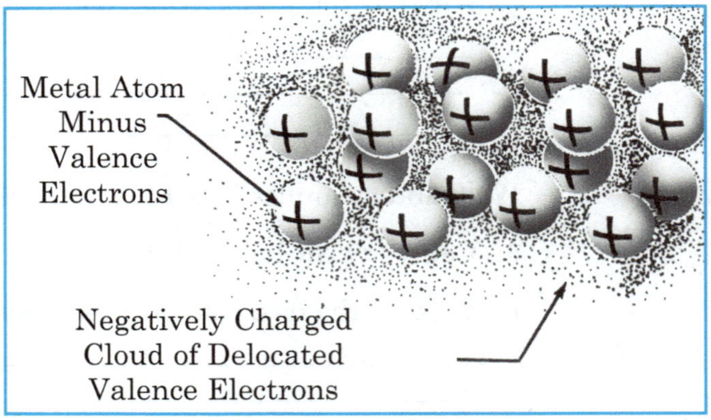

FIGURE 4: METALLIC BOND

Do these activities.

2.23 Using Figure 4 as a model for metallic bonding, try to explain why metals are such good conductors of heat and electricity.

____

____

____

____

____

____

2.24 List the predominant type of bonding as ionic, covalent, or metallic for each of the following chemicals.

a. NaCl - ____________________

b. Al rod - ____________________

c. H_2O - ____________________

d. CsF - ____________________

e. $FeCl_3$ - ____________________

f. AgI - ____________________

g. SO_2 - ____________________

h. CO_2 - ____________________

Adult check ____________________

Initial Date

Review the material in this section in preparation for the Self Test. The Self Test will check your mastery of this particular section as well as your knowledge of the previous section.

SELF TEST 2

Match these items (each answer, 2 points). Each answer may be used more than once.

2.01	_____	nondirected bond	a. metal family
2.02	_____	nonreactive (family)	b. nonmetal family
2.03	_____	highly directed bond	c. inert gas family
2.04	_____	bond with mobile electrons	d. covalent bond
2.05	_____	good conductor of electricity	e. ionic bond
2.06	_____	has low electronegativity (family)	f. metallic bond
2.07	_____	low ionization energy (family)	
2.08	_____	generally solid in pure state (family)	
2.09	_____	sharing bond	
2.010	_____	bond made up of ions	

Choose the correct answer (each answer, 3 points).

2.011 The number of moles of Cl^- ions needed to combine completely with 0.25 moles of Mg^{+2} ions is ________ .

a. 0.5 b. 0.125 c. 0.75 d. 0.25

2.012 Based upon the ion charges as taken from the Periodic Table, the formula for rubidium (Rb) chloride (Cl) is ________ .

a. $RbCl_2$ b. Rb_2Cl c. RbCl d. Rb_3Cl_2

2.013 Based upon location in the Periodic Table, the formula for beryllium (Be) chloride (Cl) is ________ .

a. Be_2Cl b. Be_2Cl_3 c. $BeCl_2$ d. Be_3Cl_4

2.014 Ionization energy and electronegativity are related to the Periodic Table in that ________ .

a. both increase going across a row and up a column
b. both decrease going across a row and increase going down a column
c. electronegativity increases going across a row and up a column while ionization energy decreases
d. there is no relation

2.015 The amount of ionic character is related to the state of matter in that ________ .

a. the more ionic the bond, the greater the tendency to be a solid
b. the more ionic the bond, the greater the tendency to be a gas
c. the more ionic the bond, the greater the tendency to be a liquid
d. there is no relation

2.016 The valence atomic orbital representation of the elements of Column IVA is ________ .

a. s^2 b. s^2p^1 c. s^2p^3 d. s^2p^2

2.017 The most active metals on the Periodic Table are found on the ________ .

a. lower left c. upper right e. middle
b. upper left d. lower right

2.018 Elements which have *high* electronegativities occur in Group ________ .

a. IA b. IIIA c. VIIA d. IIA e. O

2.019 The compound with the greatest ionic character is ________ .

a. H_2 b. NaCl c. HCF_3 d. CH_4 e. $AlCl_3$

2.020 As the atomic number of the elements in Column IIA increases, the size of the atoms generally ________ .

a. increases
b. decreases
c. stays the same
d. varies
e. cannot tell

Make the following calculations (each answer, 3 points).

2.021 Using the Periodic Table determine the percent of ionic character for each of the following compounds.

a. NaCl = __________ %
b. CO_2 = __________ %
c. H_2O = __________ %
d. KI = __________ %
e. O_2 = __________ %
f. CH_4 = __________ %

Answer the following questions (each answer, 5 points).

2.022 Explain in detail the following types of bonding.

a. covalent bonds: ____

b. ionic bonds: ____

c. metallic bonds: ____

66/83

Score
Adult check ______________
Initial **Date**

III. MOLECULAR SHAPE AND ELECTRON DISTRIBUTION

Every architect has a plan, an idea, a concept of what his creation will look like. A composer or artist has a mental image of what the finished product will be like. The Master Architect of our universe had His master plan, too. In this section you will explore the Creator's master plan concerning the architecture of molecules. Each molecule has a special shape and design. This design is not by chance; but in a regular, periodic fashion a regularity and order exists to the design of all molecules. Our task is to explore that regularity of design.

SECTION OBJECTIVES

Review these objectives. When you have completed this section, you should be able to:

5. Describe and explain the concept of hybridization.
6. Predict molecular shape by knowing the molecular formula.
7. Describe polarity and predict the polarity in molecules.

VOCABULARY

Study these words to enhance your learning success in this section.

hybrid bond
polar
polarity
hybridization

Do the following investigation.

These supplies are needed:

acetate strip and tissue paper
vinyl strip and woolen cloth
slow, steady stream of water from a faucet

Follow these directions and answer these questions. Put a check in the box when each step is completed.

❑ 1. Rub the acetate strip with the tissue paper.

❑ 2. Bring the strip near a slow stream of running water.

3.1 What happened when the strip was brought near the stream?

3.2 From your ideas about charges on acetate and vinyl strips, predict what will happen if a charged vinyl strip is brought near the slow stream of water. (Review Science LIFEPAC 1104.)

❑ 3. Test your hypothesis. Rub a vinyl strip with a woolen cloth.

❑ 4. Bring the charged strip near a slow stream of water.

SCIENCE

1105

LIFEPAC TEST

79
99

Name ______________________

Date ______________________

Score ______________________

SCIENCE 1105: LIFEPAC TEST

Match these items (each answer, 2 points).

		Elements forming compound	Formulas	
1.	_______	sodium, oxygen	a. Na_2O	b. NaO_2
2.	_______	magnesium, oxygen	c. MgO	d. Mg_2O
3.	_______	cesium, fluorine	e. CsF_2	f. CsF
4.	_______	aluminum, oxygen	g. Al_3O_2	h. Al_2O_3
5.	_______	boron, chlorine	i. BCl_3	j. BCl_2
6.	_______	tin, iodine	k. SnI_4	l. Sn_2I_3
7.	_______	barium, sulfur	m. BaS	n. BaS_2
8.	_______	gallium, oxygen	o. Ga_2O_3	p. Ga_3O_2
9.	_______	nitrogen, hydrogen	q. NH_3	r. N_2H_3
10.	_______	potassium, chlorine	s. KCl	t. K_2Cl

Circle the letter of the best response (each answer, 3 points).

11. From its position in the Periodic Table one would predict that the element potassium has the following property:
    a. has less attraction for its outer electron than does rubidium.
    b. would react with such elements as sulfur or iodine to form covalent compounds.
    c. enters into chemical reactions by taking electrons from other elements.
    d. more active than lithium.

12. The most polar of the following molecules would be
    a. CCl_4 c. CH_4 e. none—all the same polarity
    b. CH_3F d. CH_3Cl

13. Which four of the following compounds will be polar?
    a. H_2 c. $CHCl_3$ e. $AlCl_3$ g. $BeCl_2$
    b. CCl_4 d. H_2SO_4 f. H_2O h. CH_2Cl_2

14. In the Periodic Table the most active nonmetals will be found in the
    a. lower left. b. lower right. c. upper right. d. upper left.

15. When atoms from Column IA combine with atoms from Column VIIA,
    a. a nearly 100% covalent bond forms.
    b. a weak bond will form.
    c. a great deal of energy is absorbed.
    d. the resulting compounds will be very unstable.
    e. the bond will be ionic.

16. Hybrid formation has likely occurred in the molecule,
    a. BrCl. b. HCl. c. NH_3. d. $AlCl_3$.

17. If atoms with similar electronegativities form a bond, it would most likely
    a. be a polar bond. d. be a metallic bond.
    b. be a covalent bond. e. be a weak bond.
    c. be an ionic bond.

18. Although orbital notation describing element Q is $1s^2 2s^2 2p^1$, Q is found to form a volatile hybrid, QH_3, which is planar triangular in shape with all the hydrogens alike. This shape can be interpreted as due to the formation of hybrid bonds of the type

a. sp^3. b. s^2p. c. sp^2. d. s^2d. e. spd.

19. Which of the following statements concerning the nature of a chemical bond is false?

a. A chemical bond represents a balance between attractive and repulsive forces.
b. Metallic bonds are highly directed covalent bonds.
c. Covalent bonds involve the sharing of bonding parts of electrons.
d. Chemical bonds frequently have both covalent and ionic character.
e. All chemical bonds form because electrons are attracted simultaneously by two positive nuclei.

20. The most vigorous chemical reaction would likely come between

a. Fr and F. b. Ca and F. c. Na and F. d. Ba and F.

21. A covalent bond is formed by the following process:

a. gaining or losing two or more electrons by an atom.
b. sharing two electrons by two atoms.
c. sharing of an electron by two atoms.
d. loss of an electron by one atom to another.

22. The bond between F and Cl would be

a. ionic. b. pure covalent. c. metallic. d. polar covalent.

23. The substance which would be the poorest conductor of electricity at room temperature is

a. Al. b. Au. c. Co. d. S.

24. Which molecule would most likely be nonpolar?

a. I — Br

b. 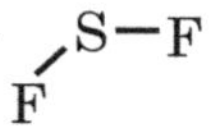

c. 

d. 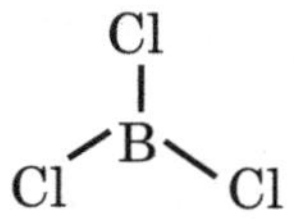

e. Br — Be — F

Make the following calculations (each answer, 3 points).

25. Using the Periodic Table of the Elements, determine the percent *ionic* character of the bond in each of the following compounds.

a. HBr = ______
b. CO_2 = ______
c. SeF_2 = ______
d. NO_2 = ______
e. IF = ______
f. HI = ______

Match these items (each answer 2 points).

26. ______ CH_4 — a. linear
27. ______ PCl_3 — b. pyramidal
28. ______ $BeBr_2$ — c. tetrahedral
29. ______ H_2O — d. bent
30. ______ BF_3 — e. planar triangular

NOTES

3.3 Now analyze your prediction.

a. Was your prediction in 3.2 correct? ________________

b. If not, propose a reason for the difference. ________________

____

____

____

3.4 Analyze the bonds of H_2O.

a. What is the percent of ionic character for the H - O bond in H_2O? ________________

b. Does this percentage mean that the H - O bond is ionic or covalent?

____

3.5 Analyze the bonds of CCl_4.

a. What is the percent of ionic character for the C - Cl bond in carbon tetrachloride, CCl_4?

____

b. Does this percentage mean that the C - Cl bond is ionic or covalent? ________________

3.6 Assume that the water stream is replaced by a stream of CCl_4. Predict what would happen in each case.

a. charged acetate strip: ____

____

____

b. charged vinyl strip: ____

____

____

c. Explain your predictions. ____

____

____

3.7 Chemists have found that charged acetate and vinyl strips have no effect on a stream of CCl_4. Does this fact match your prediction? ________________

3.8 Develop a model or picture of water, H_2O, and carbon tetrachloride, CCl_4, that would account for observations with the charged strips for the two compounds.

____

____

____

____

Adult check ________________
Initial Date

HYBRIDIZATION

Many times scientists develop models to help them answer questions about the world. In an attempt to make a complex set of variables fit a man-made model, they are unable to fit new facts and events into the old model. We then have three options: Disregard the new evidence, modify the old model to fit the new evidence, or disregard the old model and develop a new one. As a result of man's attempt to understand the laws of the universe, ideas frequently change.

In this section you will study a model that does not fit the predicted pattern. You will explore the reasons why the evidence does not fit and develop a revised model to help explain the evidence.

Column IA. Probably question 3.8 was very difficult to answer. You may have not been able to propose any model. Let us investigate a possibility together and try to develop a suitable model. The amount of electronegativity difference between the H and O in water is 1.4. This difference indicates a 61% covalent bond. The electronegativity difference between C and Cl in CCl_4 is 0.5. This difference indicates a 96% covalent bond. Therefore, the bonds in water are similar to those in carbon tetrachloride.

Since the bond character in H_2O and CCl_4 is very similar, something else must be causing the observations with charged strips and the streams of H_2O and CCl_4. One assumption that has been made up to this point is that the molecules are exactly symmetrical in arrangement. In Science LIFEPAC 1104 you explored in some depth the geometry of atoms as separate, individual particles. Now you will explore the geometry of ions and molecules to see if their geometry will lend any help in the explanation to 3.8. In considering the shape of molecules or ions, the following guidelines should be kept in mind:

1. *Only valence electrons are involved in bonding.*
2. *Only single unpaired electrons can form a bond.*

Do this activity.

3.9 Complete the first four sections of Data Table 1 on page 23.

Adult check ____________
Initial Date

Column IA elements have one valence electron. With the exception of hydrogen, all the elements are metals. Study Figure 5. Notice that each element has one valence electron unpaired and capable of bonding.

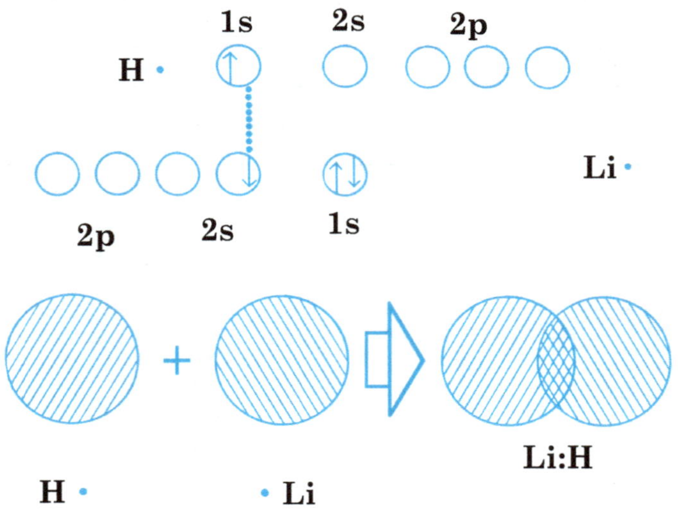

FIGURE 5

DATA TABLE 1

Activity Number	GENERAL CHARACTERISTICS OF *s* AND *p* ELEMENTS								
	COLUMN	IA	IIA	IIIA	IVA	VA	VIA	VIIA	GASES INERT
3.9	ATOMIC ORBITAL REPRESENTATION		s^2			s^2p^3			
	p	000	000	000	000	ØØØ	000	000	000
	s	0	⊗	0	0	⊗	0	0	0
3.9	TOTAL NUMBER OF VALENCE ELECTRONS		2			5			
3.9	APPARENT NUMBER OF NONBONDING ELECTRON PAIRS		1			1			4
3.9	APPARENT NUMBER OF BONDING ELECTRONS		0			3			0
3.20	ACTUAL NUMBER OF BONDING ELECTRONS	1			4				
3.20	HYBRID TYPE (IF ANY)	NONE			sp^3				
3.35	PICTURE	LiH	BeH_2	BH_3	CH_4	NH_3 	H_2O	HF	NONE
	GEOMETRY FOR THE REPRESENTATIVE COMPOUNDS. SHOW WHICH ARE BONDING ELECTRONS AND WHICH ARE PAIRED ELECTRONS. LET X REPRESENT COLUMN AND • REPRESENT COLUMN ELEMENT VALENCE ELECTRONS DOT MODEL								
		LINEAR				PYRA-MIDAL			MONATOMIC
3.36	FORMS POLAR OR NONPOLAR COM-POUND WITH HYDROGEN								

Column IIA. Notice that according to Guideline 2 stated previously, Column IIA should not combine with other elements since the elements of this column have *no unpaired* electrons available for bond formation. However, chemists have discovered compounds such as MgH_2, CaH_2, and BaH_2. How are these compounds possible?

Do these activities.

3.10 How many unpaired electrons are available in one hydrogen atom for bonding with other atoms?

____

3.11 Therefore, in order to satisfy two hydrogen atoms, how many *unpaired* electrons must each Column IIA element have available for bonding? _______________

3.12 How many valence electrons do the members of Column IIA have? _______________

3.13 What must happen to this electron pair in order for it to bond with the two hydrogen atoms?

____

____

The characteristic evidenced in Column IIA of elements unpairing already paired electrons in order to form bonds is called **hybridization**. The word hybrid, such as hybrid plants or hybrid animals, means to be a mixture of two different sources. In chemistry a **hybrid bond** is a mixture of different types of orbitals. The particular hybrid depends upon the atomic families involved. Hybridization occurs only if a vacant orbital of a very similar energy level is available to take this unpaired electron. The new hybrid bond has an energy content which is a compromise between the original level and the new level. For example, in MgH_2, the paired s^2 electrons of the Mg atom were split to give two single unpaired electrons. Since this unpairing took energy, the new bonds that formed must be made from higher energy electrons. The new bond is going to be an energy compromise between the old *s* level and the nearest vacant orbital. In the case of the IIA family, the next higher vacant orbital is a *p* orbital. This new bond can be represented in this way.

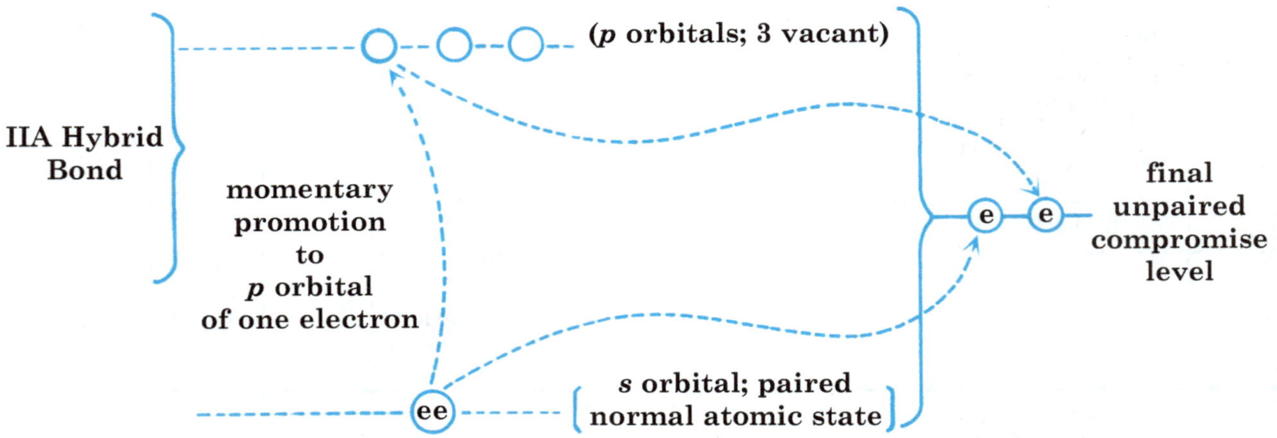

The IIA hybrid type involves two electrons. The final state is a compromise between the old *s* energy level and the promoted *p* orbital. The IIA hybrid is called an *sp hybrid* since one *s* orbital and one *p* orbital are involved.

The bonds between Be and H can be represented in this way.

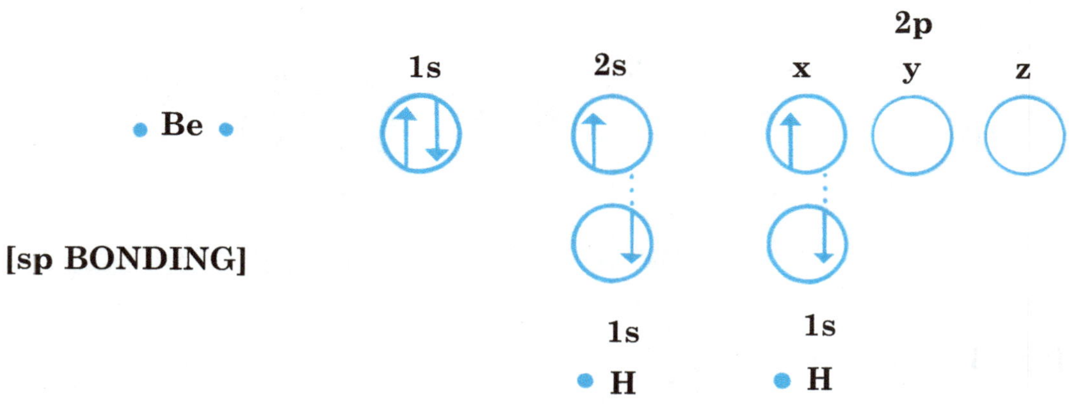

Column IIIA. A similar kind of problem of hybrid occurs with BH_3. Use the ideas of hybridization to complete the following activity.

Do these activities.

3.14 Analyze the boron atom.

a. How many valence electrons does B have? ________________

b. How many unpaired electrons are available for bond formation? ________________

A diagram like the following one might be used to explain compounds like BH_3.

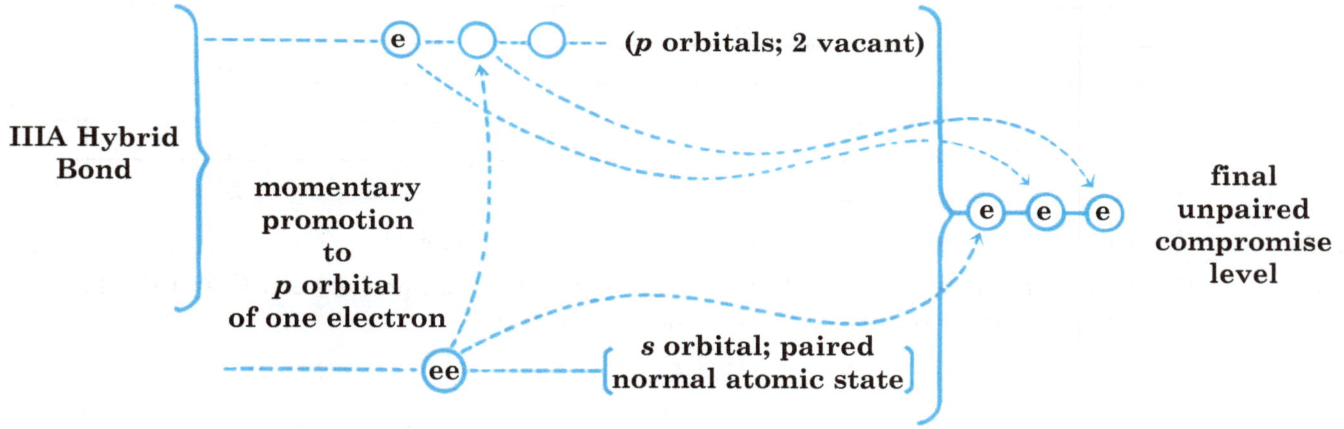

Answer this question.

3.15 The hybrid formed between IIIA elements and hydrogen is called sp^2. Why?

____

____

____

Adult check ________________
Initial Date

The bonds between B and H could be represented in this way.

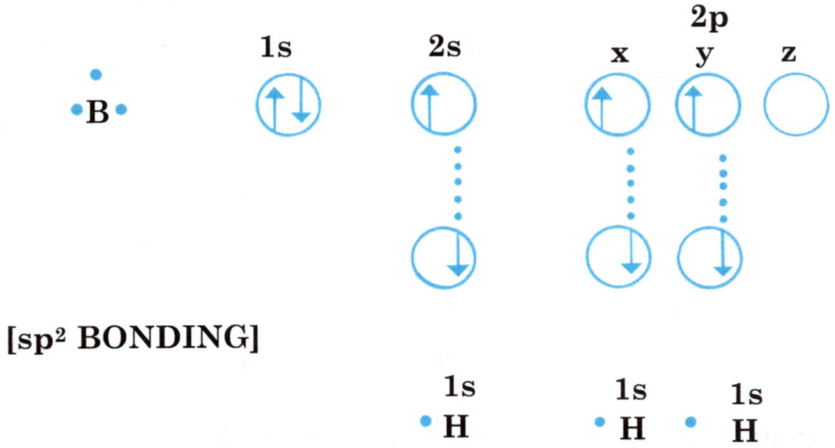

Column IVA. This column contains the common elements of carbon, silicon, tin, and lead. Carbon is the element that forms the basis of all life. Our Creator in His wisdom gave carbon unique characteristics to make life possible. Science LIFEPAC 1108 will be a detailed study of carbon and carbon compounds.

Do the following activities.

3.16 On the basis of hybrid formation, explain the existence of compounds of Column IVA such as CH_4 when the elements of Column IVA only have two unpaired electrons in the normal state.

____
____
____
____
____
____
____

3.17 What would you suggest as a name to call this type of hybrid? ____________________

3.18 Complete the following circle diagram for the bond formation between C and the four H's. Draw in the arrows representing the valence electrons.

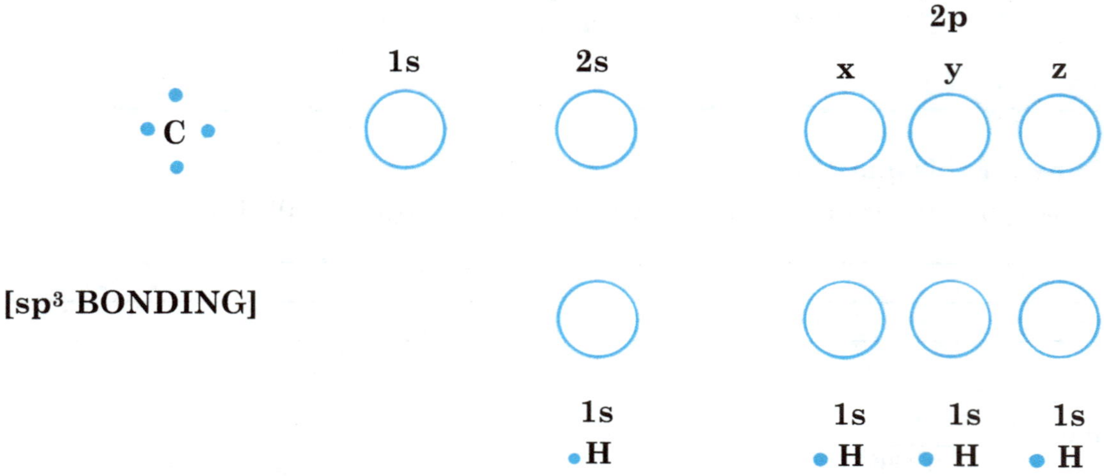

3.19 Why is it impossible to have a hybrid like sp^4, sp^5, or sp^6?

____

____

____

____

____

____

____

____

3.20 Complete Sections 5 and 6 in Data Table 1.

Adult check ____________________
Initial Date

Column VA. Elements in the columns to the right of Column IVA do not form hybrids because there is no open orbital available for the promotion of one of the paired *s* electrons.

The name given to the bond type Column VA is p^3 since the three single *p* electrons are the bonding electrons. The p^3 bonding can be represented in this way.

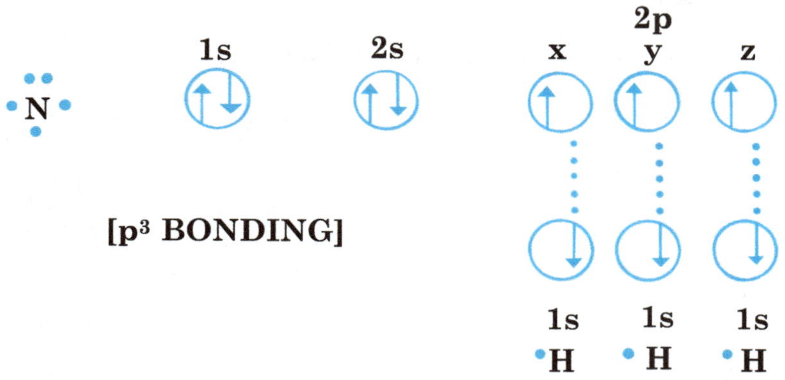

Common compounds of Column VA are ammonia (NH_3), phosphorus trichloride (PCl_3), and bismuth trichloride ($BiCl_3$).

Column VIA. Review the valence orbital structure for Column VIA. Notice that there are only two unpaired valence electrons. These two electrons are both in *p* orbitals. Bonding of Column VIA elements with hydrogen can be shown by using the example of oxygen and hydrogen in H_2O. The bond type is called p^2 bonding.

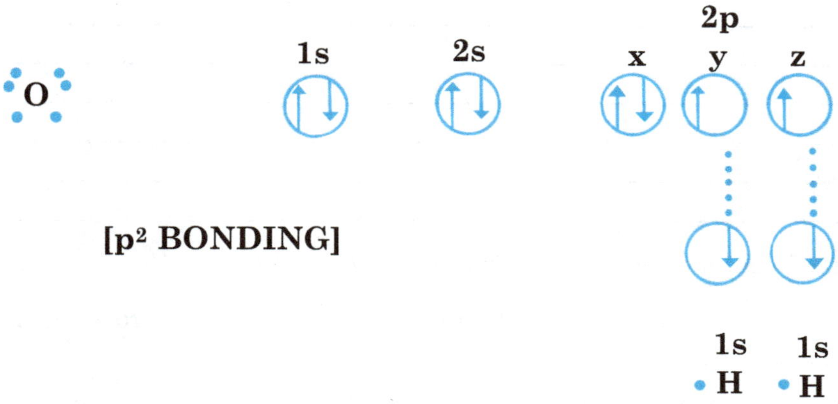

Do these activities.

3.21 Why would Column VIA bonds be called p^2 bonds?

____

____

____

3.22 Using the H_2O diagram as a model, represent the bonding of hydrogen to sulfur.

Column VIIA. Column VIIA is known as the *halogen* family. This family has a high tendency to gain electrons. The most common compound of a halogen is probably sodium chloride (table salt), NaCl.

Do these activities.

3.23 Describe in detail the following diagram. Be specific.

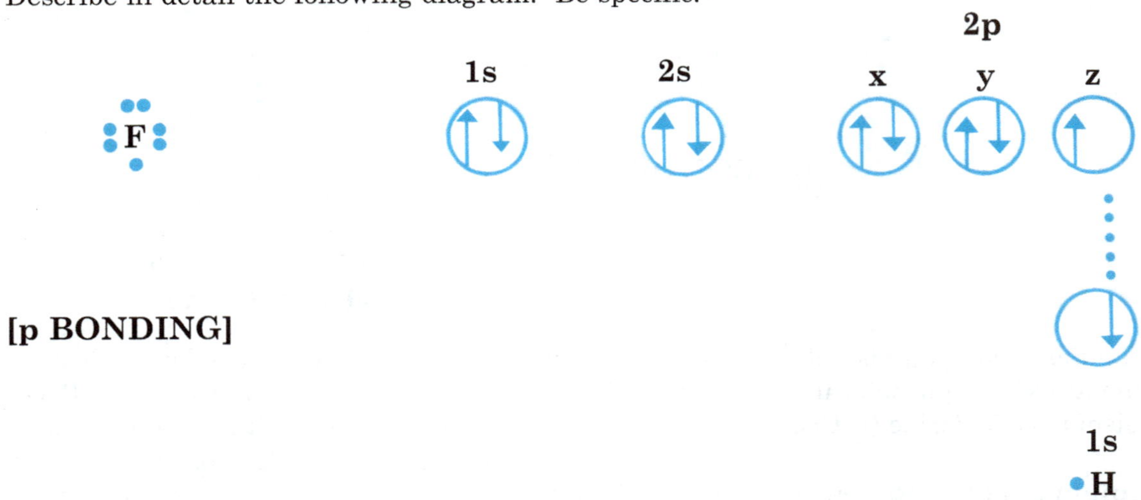

____

____

____

____

____

____

Predict the formula of halogen compounds.

3.24 a. What is the formula for the compound formed between hydrogen and fluorine?

____

b. Letting *X* represent any of the Column VIIA elements what would be the general formula of Column VIIA elements with hydrogen? ______________

Column VIIIA. The far right column of the Periodic Table is the *inert gas* family. This family gets its name from its chemical inertness. No member of this family reacts under normal laboratory conditions with any other element. Some inert gas compounds have been formed under unique laboratory conditions, but the resulting compound is very unstable and if left alone will revert back to the elements from which the compound was made. Two examples are XeF_6 and XeF_4.

Two inert gases are represented in the following diagram.

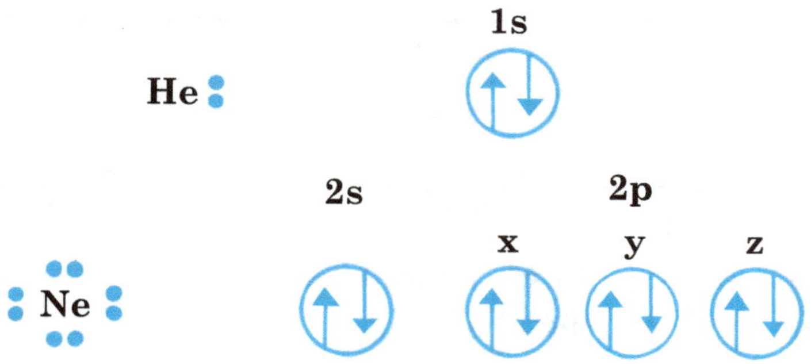

Do this activity.

3.25 Look at the electron configuration of the inert gases. Using the two guidelines for bond formation given previously in this section, explain why inert gases are inert.

____

____

____

____

____

____

____

____

____

3.26 Write a report on one of the families of elements. Include the sources, abundances, uses, and properties. Attach the report to this LIFEPAC for evaluation.

Adult check ____________________
Initial Date

POLARITY

In a hybrid the bonding electrons are *all* at the same energy level. Since these bonding electrons are at the same level they must have the same probability of being the same distance from the nucleus. We can represent this fact by saying that the bonding electrons are located on the surface of a sphere with the nucleus as the center. Since all electrons have like electrical charges, electrons tend to repel each other. If there is more than one bonding electron located on the surface of this hybrid sphere, the electrons will repel each other. Therefore, the unpaired bonding electrons try to spread out as much as possible on the surface of the sphere.

Molecular shape. As was stated previously in the LIFEPAC, the assumption has been made that all molecules are symmetrical. This assumption means that the electrons are evenly distributed in all directions. If this assumption is true, all molecules should behave much like inert gases because inert gases have a complete, symmetrical distribution of electrons within each atom. You found earlier in this LIFEPAC that H_2O and CCl_4 react much differently to charged plastic strips. We will now focus on the idea of molecular symmetry to see if we can find a significant difference in shape between H_2O and CCl_4 that might show that the electron distribution in these two compounds is greatly different. If we can discover an electron distribution difference, this difference will explain why H_2O and CCl_4 react differently to charged objects.

Do this investigation.

These supplies are needed:

one large styrofoam ball (at least a 10 cm diameter)

8 small nails about 4-5 cm long

Follow these directions and answer these questions. Put a check in each box when you have completed the step.

❑ 1. Secure the styrofoam ball and nails.

❑ 2. Place one nail at any point into the ball so the point of the nail is toward the center of the ball. The nail represents the one valence electron of Column IA and the ball represents any member of that column.

3.27 Imagine that a hydrogen atom attaches itself to the one valence electron (the nail) of the Column IA element (the ball). Describe the geometric shape of the Column IA – hydrogen compound. (Use LiH as your example.)

❑ 3. Secure a second electron (nail). You now have two valence electrons.

❑ 4. Place the second nail into the ball at the point you predict will represent the distribution of the two valence electrons on the surface of the Column IIA element. (Use Be as the representative of Column IIA.)

3.28 Imagine that one hydrogen attached itself to each of the valence electrons of Be. Describe the resulting geometry of the BeH_2 molecule.

The geometric model for Column IIA elements involves the sp hybrid. The two valence electrons of the column elements naturally repel each other and spread out from each other as much as possible. This spreading process means that the two electrons will be located on the opposite side from one another on the surface of the sphere. This geometry can be represented in this way.

sp HYBRID BONDING RESULTS IN LINEAR MOLECULES

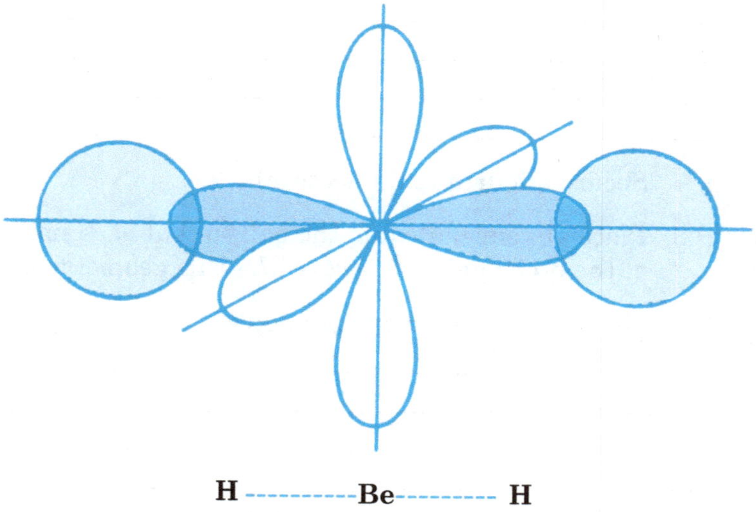

Continue the investigation. Follow the directions and answer the questions.

❑ 5. Secure a third nail (electron).

❑ 6. Place the three nails into the ball at points you think represent the location of the three valence electrons of Column IIIA elements.

3.29 Imagine that one hydrogen atom attached itself to each of the valence electrons of B. Describe the geometry of BH_3.

____

____

____

____

Study the following diagram for sp^2 bonding — the bond type for Column IIIA elements.

sp^2 HYBRID BONDING RESULTS in a PLANAR EQUILATERAL TRIANGLE

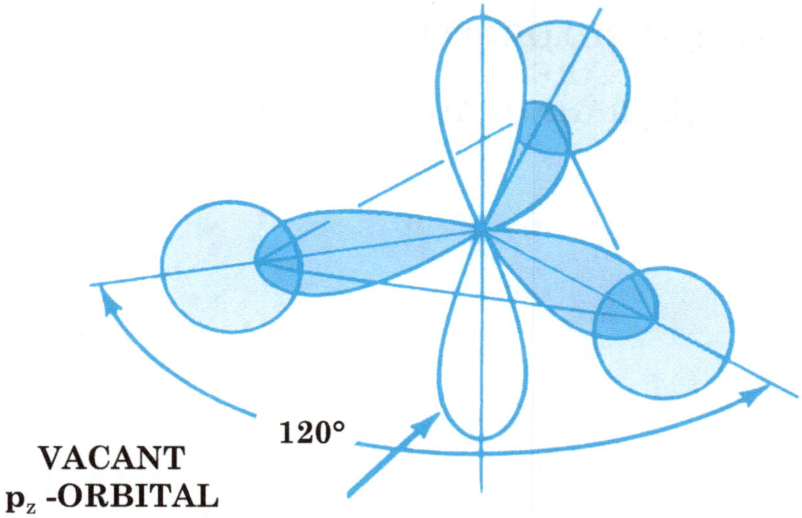

3.30 Describe the geometry of the sp^2 bond as shown and compare with what you predicted in 3.29.

____
____
____
____
____
____
____

❑ 7. Secure a fourth nail (electron).

❑ 8. Place all four nails into the ball so that each nail is located the same distance from each of the other nails. (*Be careful* – this geometry is NOT planar.)

3.31 Describe your predicted sp^3 geometry. ________________________________
____
____
____
____
____

The following diagram of the hybrid bond sp^3 shows the three-dimensional distribution of the four valence electrons for the Column IVA elements.

sp^3 HYBRID ORBITALS ARE TETRAHEDRAL

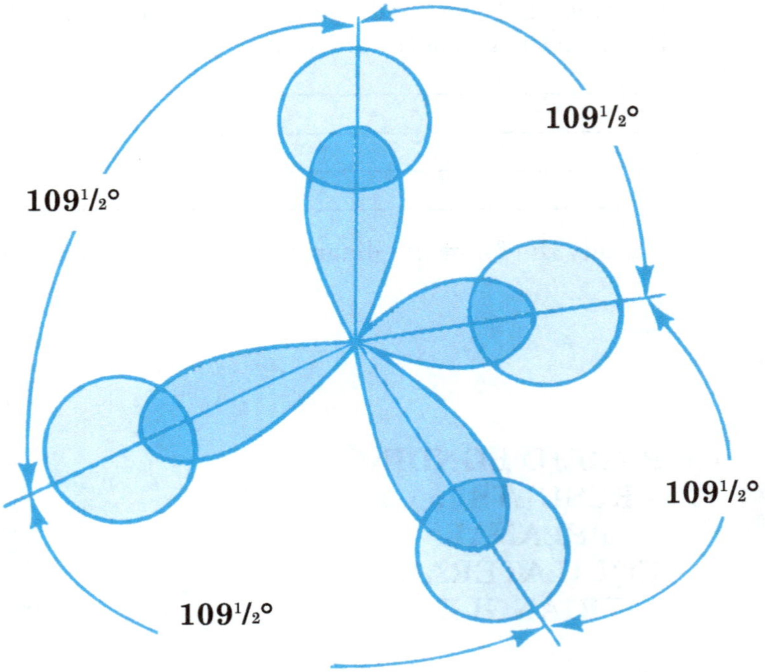

3.32 Describe the model shown and compare with your predicted geometry in 3.31.

____

____

____

____

____

____

____

❑ 9. Secure the fifth nail representing the fifth valence electron.

❑ 10. Place the fifth nail next to one of the four electrons from Step 8. You pair the fifth nail because the fifth electron is a paired electron with one already positioned on the surface of the sphere.

3.33 Study the diagrams for p^3 and p^2 bonding. Compare the geometry of each with that of the sp^3 bonding shape. Describe the geometric similarities and differences among the three bonding

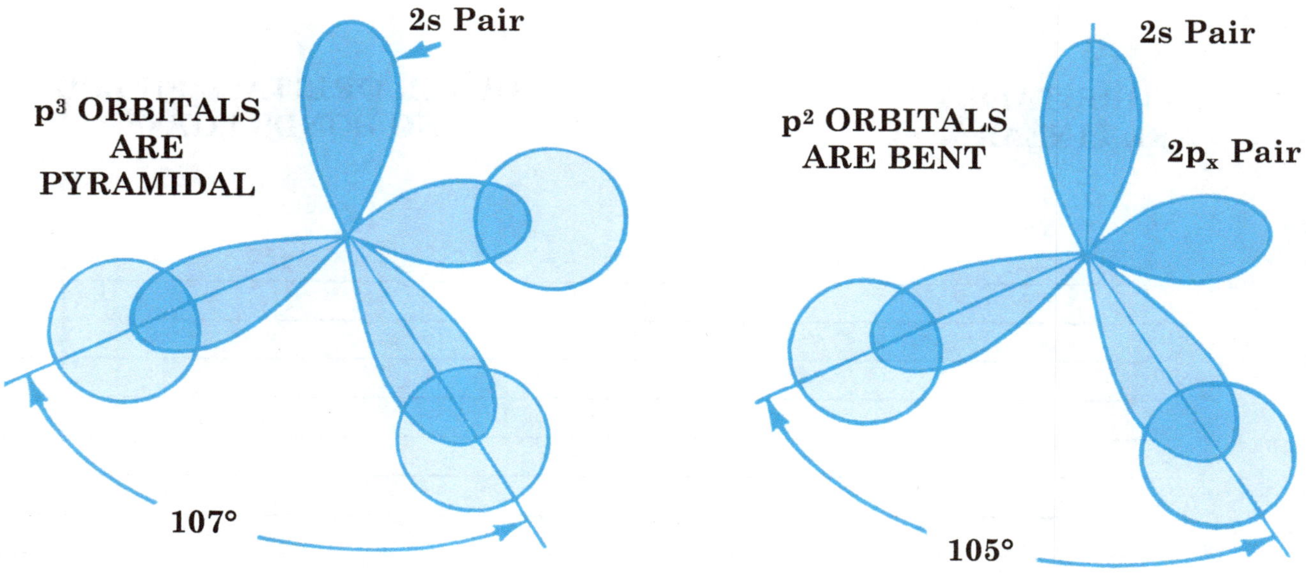

types (i.e., sp^3, p^3, and p^2). ____

____

____

____

____

____

____

____

❑ 11. Using the model from Step 10, verify the shape of Column VA and Column VIA elements.

❑ 12. Using the last two nails, place them in their proper places with the other six "electrons" located on the sphere. Notice that all electrons are paired when the eighth electron is placed in the ball.

3.34 Study the diagram for the geometry of Columns VIIA and VIIIA. Compare the geometry of each model with the geometry of the sp^3 molecular shape. Describe geometric similarities and differences among the three (i.e., sp^3, p, and inert gas) bonding types.

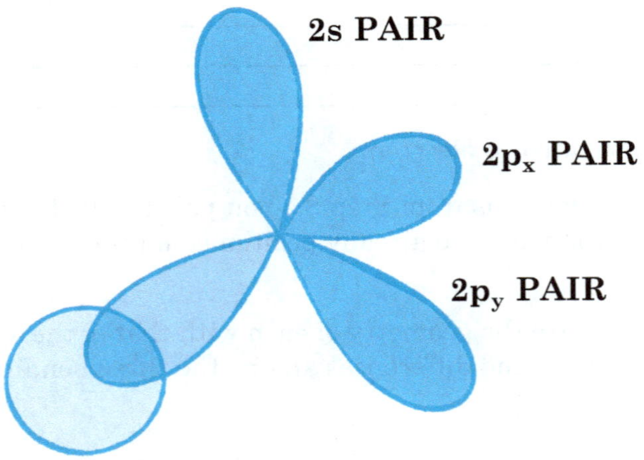

p - ORBITALS ARE LINEAR

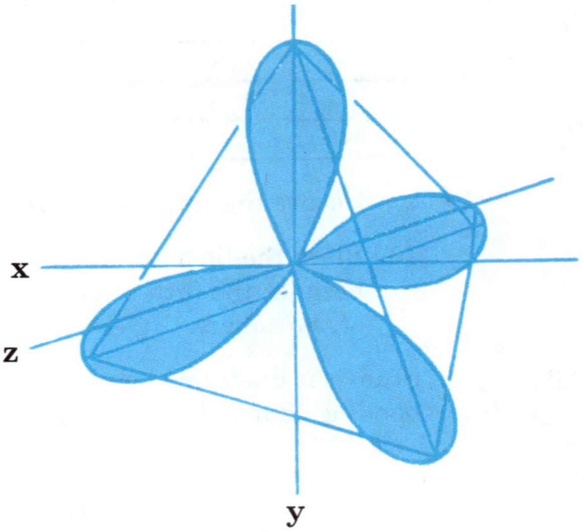

OUTER ORBITALS FILLED NO BONDS FORM

3.35 Review Activities 2.27 through 2.35. Complete Section 7 of Data Table 1.

Adult check ________________
Initial Date

A review of the geometric shapes of the molecules formed with hydrogen and the elements of the different A columns shows that the resulting molecules are not always symmetrical in shape. When a molecule is not symmetrical in shape, the electrons are not symmetrically distributed in all directions within the molecule. The molecule is electrically lopsided. When the molecule is lopsided and unbalanced, it is **polar**. A polar molecule has a positive side and a negative side. Study the following diagrams showing the symmetry and polarity of representative molecules formed from elements of each column reacted with hydrogen.

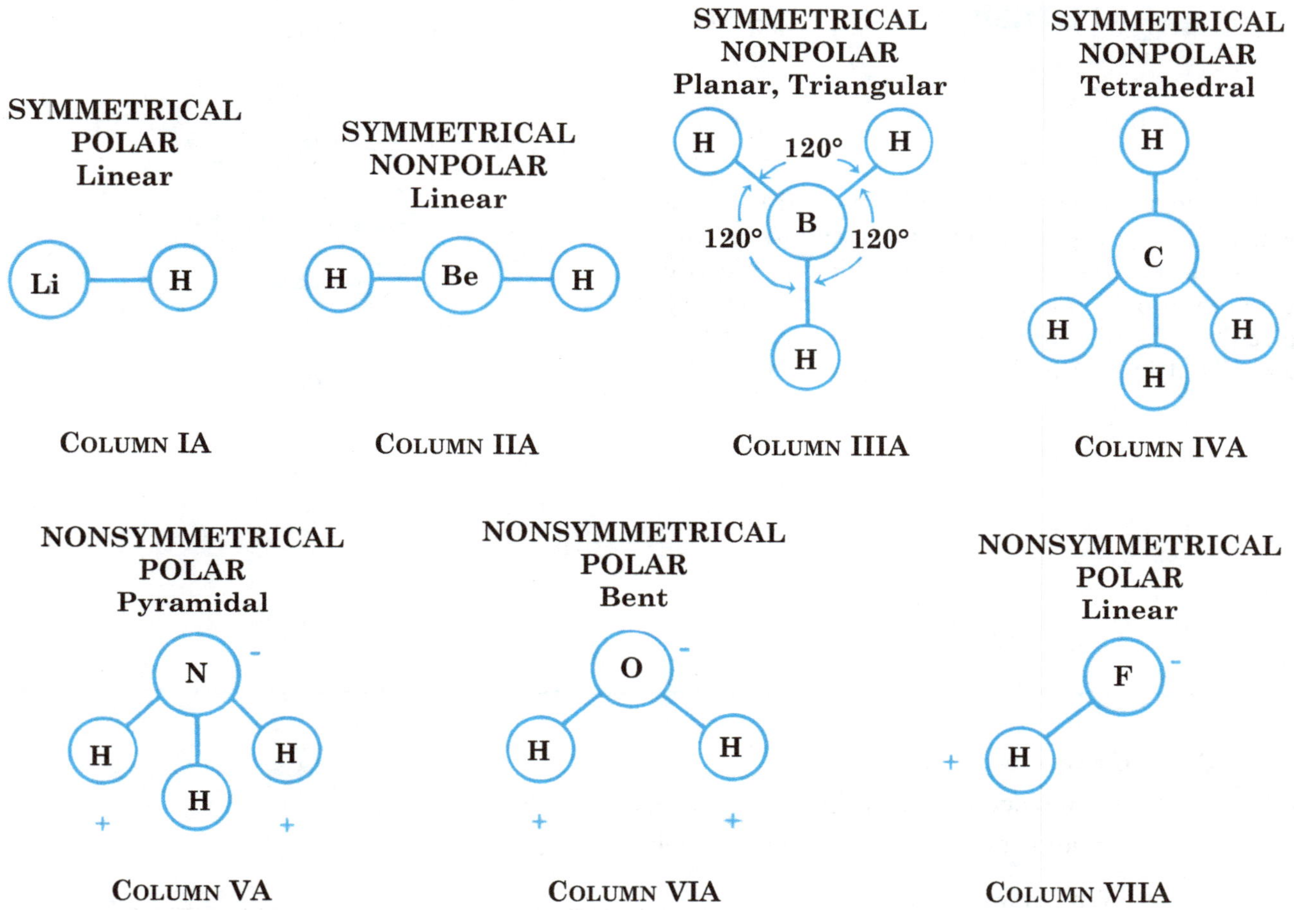

Do this activity.

3.36 Complete Section 8 of Data Table 1.

Electronegativity difference. A second factor that influences the **polarity** of a molecule is the amount of electronegativity difference between the atoms of any given bond. Even though the geometry may be symmetrical, the molecule could still be polar.

Compounds that are held together with covalent bonds often do not have atoms of equal electronegativity forming the molecules. Therefore, an unequal sharing of the bonding electrons occurs with the more electronegative atom "hogging" the bonding electrons more than 50 percent of the time. An example of this unequal sharing is LiH. Hydrogen has a desire for electrons of 2.1. Li has a value of 1.0. Since an electronegativity difference of 1.1 exists between the two atoms, this difference represents a 26 percent ionic bond characteristic. This difference means that hydrogen has more desire for the one lithium valence electron than does the lithium atom. A representation of this electron shift might be as shown in the following diagram.

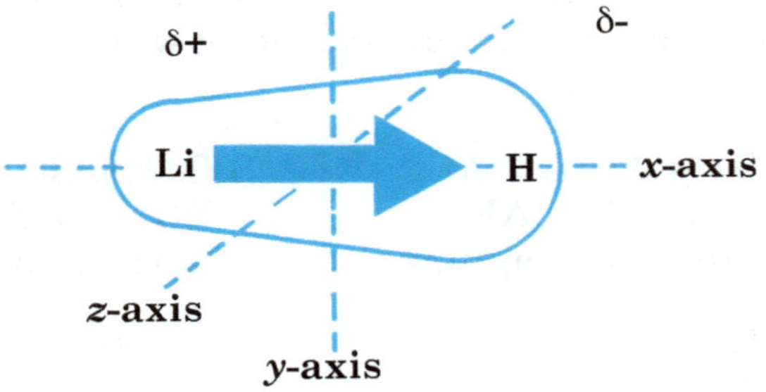

This model shows that the electrons are shifted toward the hydrogen atom and away from the lithium atom. Because of this shift of electric charge, a dipole (or two-pole) is formed in the molecule. The lithium end is partially positive (symbolized by δ+) and the hydrogen is partially negative (symbolized by δ-). Whenever an unequal distribution of electric charge exists in a molecule, the molecule is said to be polar.

An example of a symmetrical molecule is BeH_2. The pull to the left by one hydrogen is exactly offset by the pull to the right by the second hydrogen. Even though there is a small dipole in the molecule, the symmetry in each direction causes an electrical balance in the molecule.

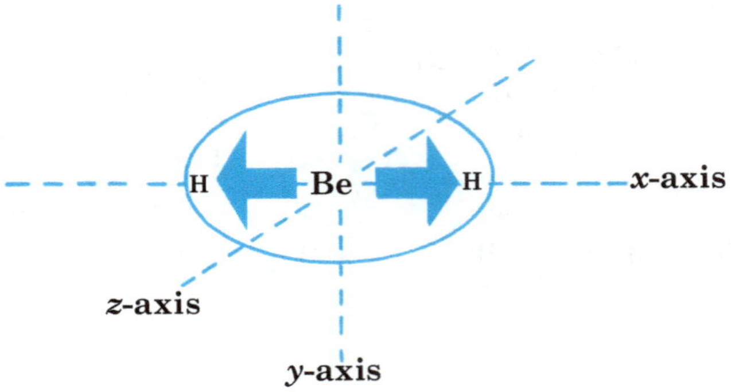

Do these activities.

3.37 Some molecules are symmetrical but still may be polar.

a. Would the molecule H-Be-Cl be polar? ____________

b. Why or why not? ____

____

____

____

____

3.38 Using the concepts of this section on polarity, describe in detail why H_2O is attracted to both a charged vinyl and a charged acetate strip while CCl_4 is not affected by either.

____

____

____

____

____

____

____

Complete the following chart using your ideas about polarity and geometric shapes of molecules and compounds.

	Compound	Geometry: Drawing	Geometry: Name of Shape	Polar and Nonpolar
3.39	H_2S	a.	b.	c.
3.40	$AlCl_3$	a.	b.	c.
3.41	HF	a.	b.	c.
3.42	CH_2Cl_2	a.	b.	c.
3.43	$SnCl_4$	a.	b.	c.
3.44	$TeCl_2$	a.	b.	c.

Complete this activity.

3.45 Look carefully at the Periodic Table. What is the geometry of the respective columns?

Column	Geometry	Column	Geometry
IA	a. ____________	VA	e. ____________
IIA	b. ____________	VIA	f. ____________
IIIA	c. ____________	VIIA	g. ____________
IVA	d. ____________	VIIIA	h. ____________

Adult check ____________________
Initial Date

OPTIONAL. Complete the following chart. This activity will be a good review of the ideas in Section III of this LIFEPAC.

	Formula	3-D Drawing	Electron-dot Formula	Basic Geometric Shape	Polarity Polar-Nonpolar
3.46	NF_3	a.	b.	c.	d.
3.47	F_2	a.	b.	c.	d.
3.48	OF_2	a.	b.	c.	d.
3.49	LiF	a.	b.	c.	d.
3.50	BeF_2	a.	b.	c.	d.
3.51	BF_3	a.	b.	c.	d.
3.52	CF_4	a.	b.	c.	d.

Review Data Table 1. Explain the table, its meaning, and concepts learned from each section to a friend. When completed get your friend's signature.

Helper check ____________________
Initial Date

Before you take this last Self Test, you may want to do one or more of these self checks.

1. _____ Read the objectives. Determine if you can do them.
2. _____ Restudy the material related to any objectives that you cannot do.
3. _____ Use the SQ3R study procedure to review the material:
   a. **S**can the sections.
   b. **Q**uestion yourself again (review the questions you wrote initially).
   c. **R**ead to answer your questions.
   d. **R**ecite the answers to yourself.
   e. **R**eview areas you didn't understand.
4. _____ Review all vocabulary, activities, and Self Tests, writing a correct answer for each wrong answer.

SELF TEST 3

Match these items (each answer, 2 points).

		Reacting elements	Possible formulas	
3.01	______	sodium-bromine	a. MgF	l. CaI_2
3.02	______	calcium-iodine	b. SiO	m. NaBr
3.03	______	aluminum-oxygen	c. Al_3O_2	n. BCl_3
3.04	______	boron-chlorine	d. Rb_2S	o. MgF_2
3.05	______	oxygen-chlorine	e. Sr_2S	p. $Ga2S_3$
3.06	______	strontium-sulfur	f. B_2Cl_3	q. SiO_2
3.07	______	magnesium-fluorine	g. Ga_3S_2	r. Al_2O_3
3.08	______	gallium-sulfur	h. O_2Cl	s. Ca_2I
3.09	______	silicon-oxygen	i. Rb_2S_3	t. Cl_2O
3.010	______	rubidium-sulfur	j. Na_2Br	u. SrS
			k. RbS_2	v. AlO

Choose the correct answer (each answer, 3 points).

3.011 From its position in the Periodic Table, one would predict that the element potassium has the following property:

a. more active than lithium.

b. enters into chemical reactions by taking electrons from other elements.

c. would react with such elements as sulfur or iodine to form covalent compounds.

d. has less attraction for its outer electron than does rubidium.

3.012 The most polar of the following molecules would be

a. CH_3F c. CH_3Cl e. all the same polarity

b. CCl_4 d. CH_4

3.013 Which five of the following compounds will be polar?

a. H_2O c. CH_2Cl_2 e. CCl_4 g. $CHCl_3$

b. HCl d. $AlCl_3$ f. H_2 h. H_2SO_4

3.014 In the Periodic Table the most active nonmetals will be found in the

a. upper right. c. upper left.

b. lower right. d. lower left.

3.015 When atoms from Column IA combine with atoms from column VIIA,

a. a nearly 100% covalent bond forms.

b. the bond will be ionic.

c. a weak bond will form.

d. the resulting compound will be very unstable.

e. a great deal of energy is absorbed.

3.016 Hybrid formation has likely occurred in the molecule,

a. $AlCl_3$. c. NH_3.

b. BrCl. d. HCl.

3.017 If atoms with similar electronegatives form a bond, it would most likely be

a. a metallic bond. d. an ionic bond.

b. a polar bond. e. a very weak bond.

c. a covalent bond.

3.018 Which of the following statements concerning the nature of a chemical bond is false?

a. All chemical bonds form because electrons are attracted simultaneously by two positive nuclei.
b. Chemical bonds frequently have both covalent and ionic character.
c. A chemical bond represents a balance between attractive and repulsive forces.
d. Metallic bonds are highly directed covalent bonds.
e. Covalent bonds involve the sharing of bonding pairs of electrons.

3.019 The most vigorous chemical reaction would likely come between

a. Na and F.
b. Ba and F.
c. Ca and F.
d. Fr and F.

3.020 A covalent bond is formed by the following process:

a. loss of an electron by one atom to another.
b. sharing of an electron by two atoms.
c. sharing two electrons by two atoms.
d. gaining or losing two or more electrons by an atom.

3.021 The bond between F and Cl would be

a. pure covalent.
b. ionic.
c. polar covalent.
d. metallic.

3.022 The substance which would be the poorest conductor of electricity at room temperature is

a. Al.
b. Au.
c. S.
d. Co.

3.023 Which molecule would most likely be nonpolar?

a. 

b. 

c. 

d. 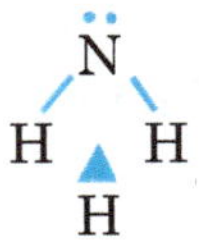

e. 

Match these items (each answer, 2 points). Each answer may be used more than once.

3.024 ________ LiI
3.025 ________ GeF_4
3.026 ________ BrF
3.027 ________ SiF_2Br_2
3.028 ________ PF_3
3.029 ________ $BeCl_2$

a. linear
b. planar triangular
c. tetrahedral
d. pyramidal
e. bent

Make the following calculations (each answer, 3 points).

3.030 Using the Periodic Table of the Elements determine the percent *ionic* character of the bond in each of the following compounds.

a. HI = ____________
b. SeF_2 = ____________
c. NO_2 = ____________
d. HBr = ____________
e. CO_2 = ____________
f. IF = ____________

Answer the following items (each answer, 5 points).

3.031 Describe an ionic bond. Use at least one example. ______________________________
____
____
____

3.032 Describe a covalent bond. Use at least one example. ______________________________
____
____
____

3.033 Describe a metallic bond. Use at least one example. ______________________________
____
____
____

83/104

Score ______________

Adult check ______________

Initial **Date**

Before taking the LIFEPAC Test, you may want to do one or more of these self checks.

1. _______ Read the objectives. Check to see if you can do them.
2. _______ Restudy the material related to any objectives that you cannot do.
3. _______ Use the SQ3R study procedure to review the material.
4. _______ Review activities Self Tests, and LIFEPAC vocabulary words.
5. _______ Restudy areas of weakness indicated by the last Self Test.

GLOSSARY

covalent bond. A chemical bond between atoms with similar electronegativities; a sharing bond; valence electrons are shared; a very directed bond between nuclei of the atoms in the bond.

electronegativity. A measure of the ability of atoms to attract (gain) electrons.

hybrid bond. A mixture of different types of orbitals.

hybridization. The process of forming a hybrid bond.

ion. Any atoms or group of atoms with unequal numbers of protons and electrons, resulting in a net ionic charge.

ionic bond. A bond between atoms of greatly differing electronegativities; a separation of one or more valence electrons from one atom (low electronegativity) toward another (high electronegativity); bonds exist between unlike electrical charges (+ and - ions).

metallic bond. The bond formed in metals, holding metals together; delocalized bond with very fluid, mobile electrons; nondirected bond with valence electrons in a "community" relationship to many nuclei; electron mobility accounts for ease of electrical conductivity.

polar. An unequal, unbalanced distribution of electrons in a molecule causing one portion to be (+) and another to be (-).

polarity. The measure of the separation and presence of a polar molecule.